AF813399

Survive The Future

Volume 3

Published by
Namya Press

NAMYA PRESS.com

India: - 213, Vardaan House, 7/28 Ansari Road Daryaganj Delhi- 110002
Email: namyapress@gmail.com
Website: - https://namyapress.com

Edition	:	First Published in 2021 (Namya Press)
Title	:	Survive the Future Vol III
Author	:	Dr. Purnima Pandey / Deepak Bishnoi
ISBN	:	978-93-5545-053-1
Copyright	:	© 2022 PPDi Institute All rights reserved

This book is being sold on the condition that it cannot be used commercially or in any other form without the prior written permission of the publisher. This book cannot be republished or sold or rented. And it cannot be operated among readers in book-binding or any other form. All these conditions will also apply to the buyer of the book. All rights of copyright are reserved in this context.

This book has been published with all efforts taken in making the material error-free after the consent of the author. However, the author and the publisher do not assume and hereby disclaim any liability of any part for loss, damage, or disruption caused by error or omissions.

The text matter and thoughts presented in this book are individual thoughts. Any similarity to other persons, living or dead, is coincidental and not intended by the author.

PPDi INSTITUTE

BE SMART PROFESSIONAL

Acknowledgement

We may not be living on Mars or traveling to work using jet packs, but there's no doubt the coming decade will bring many exciting technological advances. We're privileged to be living in a time where science and technology can assist us, make our lives easier and rethink the ways we go about our daily lives. The technology we're already exposed and accustomed to has paved the way for us to innovate further and this list of current and future technologies certainly have the potential to change our lives even more.

Digital technologies are currently shaping and transforming whole societies. Increasing access to data and digital technologies empower people. However, the digital divide still exists, and it plays out along different dimensions. By 2030, I envision an inclusive world where divisions have been reduced - especially the gender divide. For this to work, we need to make sure three things happen.

First, strengthening digital technologies skills and lifelong learning to include everyone, notably women and low-income individuals. Second, we will need to tackle risks like cybersecurity risks and the misuse of information. Third, we will need to use the digital technologies such as Artificial Intelligence and Machine Learning to help us addressing collective challenges like improving healthcare and curing diseases. Applying these policies will lead to better lives for all - notably women and low-income groups.

We are grateful to scholars, academicians and students for their support and sharing their ideas to our book.

Dr. Purnima Pandey

(Chairperson, PPDi Institute)

PhD (Education), M. Ed, PGDCA, M.Sc.

A versatile academician with an experience of 19+ years in Academics. Worked in various reputed institutes & universities on lead roles. An intellectual personality has 18 Research papers published in National and International Journals in her name. As a leader, proactively Conducted and presented various seminars and workshops for the overall development of teachers and students.

Deepak Bishnoi

(Director, PPDi Institute)

PGDBM (Information Technology & Management), PGD in Advertising & Media Management

Achievement-oriented Professional having 18+ years of experience in Media Planning, Buying & strategy of Various brand in Indian market. Excellence in conceiving and implementing advertising strategies that have fueled stronger brand development, market-share increase, revenue growth, new product launches. In addition, had also performed diverse roles as, Strategic Market Planning & Researcher, Competitive Brand Analyst, Marketing Communications & Media Analyst, Media Planning & Buyer specialist.

WELCOME
TO THE
FUTURE

Table of Contents

Chapter 1

Future Wave

In the future wave of discovering technologies, we all are investing in both qualitative and quantitative assets. Technology aggregation based on user experience and ecological thinking will be a new trend in the industry.

Flying cars, robot butlers, internet for everyone? There are innumerable futuristic technologies likes these that promises to change the world as we know it, in the coming years. The future of technology comprises of anything and everything that we can imagine, brought to life by innovation, to fulfill our needs, comforts, and wishes, which determine our future choices. Today, technology is present in every facet of our lives - from the moment we wake up till the time we go to bed, it assists, supports and makes our lives easier.

Over time, our evolving needs have led technological innovators to push further and introduce ground-breaking innovations that meet the demand. With the increasing use of data, more and more devices can now be connected. In the month ahead, there is not doubt that technologies like AI (Artificial Intelligence), AR (Augmented Reality), VR (Virtual Reality), etc will continue to redefine user experience.

A recent report mentioned that any individual looks at their phones for more than 2.5 hours and makes about 35,000 decisions on an average, in a single day. While that sounds like a mammoth task, AI or artificial intelligence is one of the technologies that helps us do this efficiently and effectively.

From digital assistants recognising voices and operating various smart devices as per instructions to simple object recognition

and information search, AI is intricately woven into our technological lives. AI even powers the security features like face recognition, voice recognition and fingerprint technology.

What's even more exciting is how the combination of this technology with augmented really is giving rise to a whole new dimension! This has digitally amalgamated the physical and virtual world, thus transforming the user experience.

For instance, the AR Glass which when powered by smartphones, provides users with a more immersive and holistic visual experience. Some of the other endless possibilities of applying this technology include AR multi-level player games, digitally placing furniture in our rooms, immersive virtual traveling, creating prototypes and structural designs - all through a device compact enough to fit into our pockets.

All about Screens

For smartphone makes specially, innovation is key. The screen for one, is a crucial aspect of a mobile device and the focus of innovation. From the time when mobile phones had analog screens that could only be used as a basic display, we have come a long way to an era of touch recognition technology and a screen's ability to sense their surroundings and switch on and off.

Innovation in screen technology is driven by consumer needs and usage of the device for both entertainment and work purposes. Users stream video content, play games, and also execute business operations through their smartphones, creating a need for a bigger, better screens with the same sleek, lightweight and compact body. Hence, the need to make the pictures on the display clearer, increase the refresh rate, make

devices sleeker, decrease heating over long usage, and most importantly, on innovations to make the screen even bigger.

The future of screen technology is focused on increasing the dimensions of the display for additional visual and functionality comfort. Innovative technologies are being created where screens can be compressed into two or more folds, and be opened up like a book, creating a tablet-like display. The other concept is a more futuristic design where by the screen wraps around the edges of the phone, increasing the display size.

In 2020, these concepts have actually been showcased to the world, paving the way for more futuristic designs next. It will be interesting to see what others way can be used to increase the size of the screen in order to make the experience more immersive.

Just like screen technologies, innovation in charging technology has taken precedence. To cater to the need for longer battery life and faster charging due to the increased use of smartphones, the industry introduced fast-charging technology and wireless charging of ease of use and functionality comfort.

Wireless charging brought the evolution of charging tech to the forefront. The latest wireless chargers are equipped to withstand a power of 125W and can charge an 4000mAh battery in 20 minutes at the fastest rate. We now focus on super fast charging capabilities to give users complete freedom to use their smartphone whenever and wherever.

Exploring the future with humanism as the backbone

We are at the verge of a new decade, and it will be an era of integration of all things. Where a device can talk to other devices, where a device itself has the ability to learn and where the multi-device composition of the interconnected ecology provide services to users.

The developing 5G technologies, coupled with the investments in the research and development around a multi-device ecosystem will definitely lead us to a more connected future.

But while we are still here, we should not lose the ability to empower people with technology, and continue to be people oriented, user centric with user experience driving products and technology. A large part of invention and innovations should come from understanding consumer needs, after all, consumer is king.

Technology should empower people and help them capture the beauty around them and unleash their imagination, thereby ensuring everyone gets the best out of life.

Whether you like it or not, technology is rapidly improving, offering new innovations and revolutionary projects every year. Some of the very sharpest minds are out there creating the next piece of future technology that will completely change how we live our lives. It can feel like scientific progress is steady but we have lived through a period of immense technological improvement in the last half century.

There are innovations happening right now that are ripped straight from the pages of science-fiction. Whether that is robots that can read minds, AI that can create images on their own, holograms, bionic eyes, or other mind-blowing technology, there is a lot to expect from the world of future technology. Below we've picked out some of the biggest and most interesting ideas.

AI Image-Generation

Researchers at the company Open AI have created a software that is able to create images from just worded prompts.Type in ‘a dog wearing a cowboy hat singing in the rain’ and you’ll get a host of completely original images that fit that description. You can even choose what style of art your request will come back in. However, the technology isn't perfected and still has issues, like when we gave it poor prompts on designing cartoon characters.

This technology known as Dall-E is now its second iteration and the team behind it plans to continue developing it further. In the future, we could see this technology used to create art exhibitions, for companies to get quick, original illustrations or of course, to revolutionize the way we create memes on the internet.

Brain Reading robots

No longer a science fiction trope, the use of brain reading technology has improved hugely in recent years. One of the most interesting and practical uses we’ve seen tested so far comes from researchers at the Swiss Federal Institute of Technology Lausanne (EPFL).

Thanks to a machine-learning algorithm, a robot arm and a brain-computer interface, these researchers have managed to create a means for tetraplegic patients (those who can’t move their upper or lower body) to interact with the world. In tests, the robot arm would perform simple tasks like moving around an obstacle. The algorithm would then interprets signals from the brain using an EEG cap and automatically determine when the arm had made a move that the brain considered incorrect, for example moving too close to the obstacle or going too fast.

Over time the algorithm can then adjust to the individuals preferences and brain signals. In the future this could lead to wheelchairs controlled by the brain or assistance machines for tetraplegic patients.

3D Printed Bones

3D printing is an industry promising everything from cheap house building through to affordable rugged armour, but one of the most interesting uses of the technology is the building of 3D printed bones.

The company Ossiform specialises in medical 3D printing, creating patient-specific replacements of different bones from tricalcium phosphate - a material with similar properties to human bones.

Using these 3D printed bones is surprisingly easy. A hospital can perform an MRI which is then sent to Ossiform who create a 3D model of the patient-specific implant that is needed. The surgeon accepts the design and then once it is printed, it can be used in surgery.

What is special about these 3D printed bones is that because of the use of tricalcium phosphate, the body will remodel the implants into vascularised bone. That means they will enable the full restoration of function that the bone it is replacing had. To achieve the best integration possible, the implants are of a porous structure and feature large pores and canals for cells to attach to and reform bone.

Realistic holographs

Holograms have been filling science fiction books, films and culture for years now, and while it does exist, it remains a difficult thing to achieve, especially on a large scale. However, a potential technology that could change this is holobricks. Developed by researchers from the University of Cambridge and Disney Research, holobricks are a way of tiling together multiple holograms to produce a large, seamless 3D image.

The issue with most holographs right now is the amount of data that they require to make, especially when done on a large scale. A regular HD display for a 2D image takes about 3GB per second to generate. A hologram of a similar size and resolution would be nearer to 3TB per second which is a huge amount of data.

To combat this, holobricks would provide individual sections of one large holographic image, heavily reducing the amount of data needed. This could eventually lead to the use of holograms in daily consumer entertainment like movies, games and digital displays.

Clothes that can hear

Wearable technology has come leaps and bounds over the years, adding new functionalities to the accessories and clothes we wear day to day. One promising avenue involves giving clothes ears, or at least the same capacity as an ear.

Researchers at MIT have created a fabric that is able to detect a heartbeat, hand claps or even very faint sounds. The team suggested that this could be used wearable tech for the blind, used in buildings to detect cracks or strains, or even woven into fishnets to detect the sound of fish.
For now, the material used is thick and a work in progress but they hope to roll it out for consumer use over the next few years.

Lab-made dairy products

You've heard of cultured "meat" and Wagyu steaks grown cell by cell in a laboratory, but what about other animal-based foodstuffs? A growing number of biotech companies around the world are investigating lab-made dairy, including milk, ice-cream, cheese and eggs. And more than one think they've cracked it.

The dairy industry is not environmentally friendly, not even close. It's responsible for 4 per cent of the world's carbon emissions, more than air travel and shipping combined, and demand is growing for a greener splash to pour into our tea cups and cereal bowls.

Compared with meat, milk isn't actually that difficult to create in a lab. Rather than grow it from stem cells, most researchers attempt to produce it in a process of fermentation, looking to produce the milk proteins whey and casein. Some products are already at market in the US, from companies such as Perfect Day, with ongoing work focused on reproducing the mouthfeel and nutritional benefits of regular cow's milk. Beyond that, researchers are working on lab-produced mozzarella that melts perfectly on top of a pizza, as well other cheeses and ice-cream.

Hydrogen planes

Carbon emissions are a huge concern when it comes to commercial flights, but there is a potential solution and it has received a lot of funding. A £15 million UK project has unveiled plans for a hydrogen-powered plane. This project is known as Fly Zero and is being led by the Aerospace Technology Institute in conjunction with the UK government.

The project has come up with a concept for a mid-size plane powered completely by liquid hydrogen. It would have the capacity to fly roughly 279 passengers halfway around the world without stopping. If this technology could be actualised, it could mean a zero-carbon flight with no stops between London and Western America or London to New Zealand with a single stop.

Digital "twins" that track your health

In Star Trek, where many of our ideas of future technology germinated, human beings can walk into the medbay and have their entire body digitally scanned for signs of illness and injury. Doing that in real life would, say the makers of Q Bio, improve health outcomes and alleviate the load on doctors at the same time.

The US company has built a scanner that will measure hundreds of biomarkers in around an hour, from hormone levels to the fat building up in your liver to the markers of inflammation or any number of cancers. It intends to use this data to produce a 3D digital avatar of a patient's body – known as a digital twin – that can be tracked over time and updated with each new scan. Q Bio CEO Jeff Kaditz hopes it will lead to a new era of preventative, personalised medicine in which the vast amounts of data collected not only help doctors prioritise which patients need to be seen most urgently, but also to develop more sophisticated ways of diagnosing illness.

Virtual Reality Universes

After making its dramatic name change, the company once known as Facebook has become Meta. This marks Zuckerberg and his huge team's move into the metaverse – an embodied internet mostly accessed through virtual and augmented reality. As part of this move, we will start to see Meta putting more time into equipment for accessing this new world – mostly in VR. Announced back in 2021, Meta has been developing a new headset under the title 'Project Cambria'. Unlike the brand's previous VR ventures like the Oculus Quest 2, this won't be a device for the average consumer, instead looking to offer the best VR experience they can make.

The Cambria has been reported to be focused on advanced eye and face tracking (to improve accuracy of avatars and your in-game movements), a higher resolution, increased field-of-view and even trying to make the headset significantly smaller. Between Meta, Google, Sony and plenty of other big tech companies, VR is getting lots of funding right now and will be seeing drastic improvements in the next couple of years.

Direct air Capture

Through the process of photosynthesis, trees have remained one of the best ways to reduce the levels of CO2 in the atmosphere. However, new technology could perform the same role as trees, absorbing carbon dioxide at greater levels while also taking up less land.

This technology is known as Direct Air Capture (DAC). It involves taking carbon dioxide from the air and either storing the CO2 in deep geological caves under ground, or using it in combination with hydrogen to produce synthetic fuels. While this technology has great potential, it has a lot of complications right now. There are now direct air capture facilities up and running, but the current models require a huge amount of energy to run. If the energy levels can be reduced in the future, DAC could prove to be one of the best technological advances for the future of the environment.

Green Funerals

Sustainable living is becoming a priority for individuals squaring up to the realities of the climate crisis, but what about eco-friendly dying? Death tends to be a carbon-heavy process, one last stamp of our ecological footprint. The average cremation reportedly releases 400kg of carbon dioxide into the atmosphere, for example. So what's a greener way to go?

In Washington State in the US, you could be composted instead. Bodies are laid in chambers with bark, soil, straw and other compounds that promote natural decomposition. Within 30 days, your body is reduced to soil that can be returned to a garden or woodland. Recompose, the company behind the process, claims it uses an eighth of the carbon dioxide of a cremation.

An alternative technology uses fungi. In 2019, the late actor Luke Perry was buried in a bespoke "mushroom suit" designed by a start-up called Coeio. The company claims its suit, made with mushrooms and other microorganisms that aid decomposition and neutralise toxins that are realised when a body usually decays.

Most alternative ways of disposing of our bodies after death are not based on new technology; they're just waiting for societal acceptance to catch up. Another example is alkaline hydrolysis, which involves breaking the body down into its chemical components over a six-hour process in a pressurized chamber. It's legal in a number of US states and uses fewer emissions compared with more traditional methods.

Artificial Eyes

Bionic eyes have been a mainstay of science fiction for decades, but now real-world research is beginning to catch up with far-sighted storytellers. A raft of technologies is coming to market that restore sight to people with different kinds of vision impairment. In January 2021, Israeli surgeons implanted the world's first artificial cornea into a bilaterally blind, 78-year-old man. When his bandages were removed, the patient could read and recognise family members immediately. The implant also fuses naturally to human tissue without the recipient's body rejecting it.

Likewise in 2020, Belgian scientists developed an artificial iris fitted to smart contact lenses that correct a number of vision disorders. And scientists are even working on wireless brain implants that bypass the eyes altogether. Researchers at Montash University in Australia are working on trials for a system whereby users wear a pair of glasses fitted with a camera. This sends data directly to the implant, which sits on the surface of the brain and gives the user a rudimentary sense of sight.

Airports for drones and flying taxis

Our congested cities are in desperate need of a breather and relief may come from the air as opposed to the roads. Plans for a different kind of transport hub – one for delivery drones and electric air-taxis – are becoming a reality, with the first Urban Air Port receiving funding from the UK government.

It's being built in Coventry. The hub will be a pilot scheme and hopefully a proof of concept for the company behind it. Powered completely off-grid by a hydrogen generator, the idea is to remove the need for as many delivery vans and personal cars on our roads, replacing them with a clean alternative in the form of a new type of small aircraft, with designs being developed by Hyundai and Airbus, among others.

Infrastructure is going to be important. Organisations like the Civil Aviation Authority are looking into the establishment of air corridors that might link a city centre with a local airport or distribution centre.

Energy Storing bricks

Scientists have found a way to store energy in the red bricks that are used to build houses. Researchers led by Washington University in St Louis, in Missouri, US, have developed a method that can turn the cheap and widely available building material into "smart bricks" that can store energy like a battery. Although the research is still in the proof-of-concept stage, the scientists claim that walls made of these bricks "could store a substantial amount of energy" and can "be recharged hundreds of thousands of times within an hour".

The researchers developed a method to convert red bricks into a type of energy storage device called a super capacitor. This involved putting a conducting coating, known as Pedot, onto brick samples, which then seeped through the fired bricks' porous structure, converting them into "energy storing electrodes". Iron oxide, which is the red pigment in the bricks, helped with the process, the researchers said.

Sweat powered smartwatches

Engineers at the University of Glasgow have developed a new type of flexible supercapacitor, which stores energy, replacing the electrolytes found in conventional batteries with sweat. It can be fully charged with as little as 20 micro litre of fluid and is robust enough to survive 4,000 cycles of the types of flexes and bends it might encounter in use.

The device works by coating polyester cellulose cloth in a thin layer of a polymer, which acts as the super capacitor's electrode. As the cloth absorbs its wearer's sweat, the positive and negative ions in the sweat interact with the polymer's surface, creating an electro chemical reaction which generates energy.

“Conventional batteries are cheaper and more plentiful than ever before but they are often built using unsustainable materials which are harmful to the environment," says Professor Ravinder Dahiya, head of the Bendable Electronics and Sensing Technologies (Best) group, based at the University of Glasgow’s James Watt School of Engineering.
“That makes them challenging to dispose of safely and potentially harmful in wearable devices, where a broken battery could spill toxic fluids on to skin. “What we’ve been able to do for the first time is show that human sweat provides a real opportunity to do away with those toxic materials entirely, with excellent charging and discharging performance.

Self-healing living concrete

Scientists have developed what they call living concrete by using sand, gel and bacteria. Researchers said this building material has structural load-bearing function, is capable of self-healing and is more environmentally friendly than concrete – which is the second most-consumed material on Earth after water.

The team from the University of Colorado Boulder believe their work paves the way for future building structures that could “heal their own cracks, suck up dangerous toxins from the air or even glow on command”.

Living Robots

Tiny hybrid robots made using stem cells from frog embryos could one day be used to swim around human bodies to specific areas requiring medicine, or to gather micro plastic in the oceans. “These are novel living machines,” said Joshua Bongard, a computer scientist and robotics expert at the

University of Vermont, who co-developed the millimetre-wide bots, known as xenobots. “They’re neither a traditional robot nor a known species of animal. It’s a new class of artefact: a living, programmable organism."

Internet for everyone

We can’t seem to live without the internet (how else would you read sciencefocus.com?), but still only around half the world’s population is connected. There are many reasons for this, including economic and social reasons, but for some the internet just isn’t accessible because they have no connection. Google is slowly trying to solve the problem using helium balloons to beam the internet to inaccessible areas, while Facebook has abandoned plans to do the same using drones, which means companies like Hiber are stealing a march.

They have taken a different approach by launching their own network of shoebox-sized micro-satellites into low Earth orbit, which wake up a modem plugged into your computer or device when it flies over and delivers your data. Their satellites orbit the Earth 16 times a day and are already being used by organisations like The British Antarctic Survey to provide internet access to very extreme of our planet.

Chapter 2

Future Survival In High-Tech Era

Everyone is aware of the fact that, technology is rapidly improving, offering new innovations and revolutionary projects every year. Some of the very sharpest minds are out there creating the next piece of future technology that will completely change how we live our lives. It can feel like scientific progress is steady but we have lived through a period of immense technological improvement in the last half century.

There are innovations happening right now that are ripped straight from the pages of science-fiction. Whether that is robots that can read minds, AI that can create images on their own, holograms, bionic eyes, or other mind-blowing technology, there is a lot to expect from the world of future technology.

Consumers are expecting constant innovation in the high-tech era we live in today and if companies can't keep up, the consequences aren't as forgiving. Brands aren't as resilient today as they once were and many are being forced from a highly competitive market because they lack the qualities it takes to survive. Nokia was once a leading company in hardware but fell behind as companies such as Apple and Google recognized the potential of the software. Both Apple and Google remain leaders of the technology industry because they are able to anticipate the future rather than react to what has already been done.

Years ago, it was common to connect with your customers through a basic website. However, many companies have realized that you now need to target mobile users in order to keep up with the technological advancements. If you aren't providing your customers with information right to their fingertips, they'll likely forget about your brand altogether. Mobile apps have become a lifestyle tool used to communicate, learn, shop, plan, play, and work. Personalizing your mobile app based on the user's preferences will enhance the overall user experience and increase user retention. Today, mobile devices are extensions of people's lives and if your content isn't personalized to each user, you'll risk losing a trusted connection and a loyal customer.

Surviving the Future

As we already know that technology is advancing at a very high speed i.e., we are headed into the high-tech era. So, to survive in the high-tech era, everyone must be prepared for whats to come ahead and for that people should be aware of the upcoming advancement in the technical sector. To survive the high-tech era and to catch up and compete with the world, people must learn and adapt.

There are multiple fields in the technical sector which are working day and night for the advancement of technology. Some of the fields are Artificial Intelligence (AI), Robotics, Automation, Virtual Reality (VR) and many more which are necessary and act as a useful tool. These fields not only beneficial for the advancement of technology but also can be used as a tool to learn and adapt to such technologies which will ultimately help in survival.

How to Adapt

To survive the advancing future, the first step to be taken is to adapt and learn. Below are the following key points which will help a person to adapt the advancement and how to proceed further.

1. Know Your Audience:

Steve Jobs said, "people don't know what they want until you show it to them". Knowing the market trends and foreseeing what's next to come by anticipating the needs of your audience is what will get you ahead of your competition. When you're unable to foresee future trends and demands, you won't have the time or the resources to adapt and adjust in accordance with the changes. Being a leader in the tech industry means being able to foresee how the market demands will change. For example, banking once required a lot more effort and now, many banks are realizing that users want access to banking without leaving their home to visit a branch. Companies that are able to adapt to change anticipate gaps, problems, and most importantly, customer needs in the market – always thinking about their next step.

2. Monitor Your Competition

Knowing your customers is one thing, but knowing your competition is just as significant. Understanding what's already on the market and demonstrating how your product or brand is more valuable and unique in comparison to your competition will win over potential customers. Nokia focused too much on hardware and not on what their competition was doing, which was software. There's a fine line between imitating the competition and paving your own path. It's important to be innovative while keeping a close eye on the competition.

3. Plan For The Future

Take responsibility to push new ideas and challenge existing strategies by planning for the future. Cut your losses with every setback, analyze what went wrong, and act on your next move. Resilient companies are able to seize the opportunities before they are even presented.

4. Inspire Innovation

Find new ways to inspire innovation by encouraging creativity within your team. Mock discoveries and brainstorming sessions are good ways for the team to think of new possibilities before a project is started. Compare similar products in the market and analyze its pros and cons to determine the user's pain points. Getting together with your team to inspire new ideas will help to create high-quality, innovative products. Innovation should come from actively planning and not merely imitating products that are already on the market. Stand out from the competition by determining a creative process to develop products that advance far beyond the competition.

Important Technologies

There are two important factors which are important when it comes to the case of "Adapt to Survive". These two important factors are - 'How to adapt' and 'What to adapt'. There are certain technologies which can be used as a tool to upgrade learning and adapt to the high-tech era. Learning these technologies will help a person survive the high-tech era. Some of these technologies are as follows:

1. Artificial Intelligence: As artificial intelligence continues to perform jobs just as well as humans, there is a new industry to add to the list – the world of art. Researchers at the company Open-AI have created a software that is able to create images from just worded prompts.

Type in 'a dog wearing a cowboy hat singing in the rain' and you'll get a host of completely original images that fit that description. You can even choose what style of art your request will come back in. However, the technology isn't perfected and still has issues, like when we gave it poor prompts on designing cartoon characters.

This technology known as Dall-E is now its second iteration and the team behind it plans to continue developing it further. In the future, we could see this technology used to create art exhibitions, for companies to get quick, original illustrations or of course, to revolutionize the way we create memes on the internet.

2. Robotics: Robotics is the intersection of science, engineering and technology that produces machines, called robots, that substitute for (or replicate) human actions. Pop culture has always been fascinated with robots. R2-D2. Optimus Prime. WALL-E. These over-exaggerated, humanoid concepts of robots usually seem like a caricature of the real thing or are they more forward thinking than we realize? Robots are gaining intellectual and mechanical capabilities that don't put the possibility of a R2-D2-like machine out of reach in the future.

As technology progresses, so too does the scope of what is considered robotics. In 2005, 90% of all robots could be found assembling cars in automotive factories. These robots consist mainly of mechanical arms tasked with welding or screwing on certain parts of a car. Today, we're seeing an evolved and expanded definition of robotics that includes the development, creation and use of bots that explore Earth's harshest conditions, robots that assist law-enforcement and even robots that assist in almost every facet of healthcare.

3. 3D Printing: 3D printing is an industry promising everything from cheap house building through to affordable rugged armor, but one of the most interesting uses of the technology is the building of 3D printed bones.

The company Ossiform specializes in medical 3D printing, creating patient-specific replacements of different bones from tricalcium phosphate – a material with similar properties to human bones.

Using these 3D printed bones is surprisingly easy. A hospital can perform an MRI which is then sent to Ossiform who create a 3D model of the patient-specific implant that is needed. The surgeon accepts the design and then once it is printed, it can be used in surgery.

What is special about these 3D printed bones is that because of the use of tricalcium phosphate, the body will remodel the implants into vascularized bone. That means they will enable the full restoration of function that the bone it is replacing had. To achieve the best integration possible, the implants are of a porous structure and feature large pores and canals for cells to attach to and reform bone.

4. Realistic Holographs: Holograms have been filling science fiction books, films and culture for years now, and while it does exist, it remains a difficult thing to achieve, especially on a large scale. However, a potential technology that could change this is holobricks.
Developed by researchers from the University of Cambridge and Disney Research, holobricks are a way of tiling together multiple holograms to produce a large, seamless 3D image.

The issue with most holographs right now is the amount of data that they require to make, especially when done on a large scale.

A regular HD display for a 2D image takes about 3GB per second to generate. A hologram of a similar size and resolution would be nearer to 3TB per second which is a huge amount of data.

To combat this, holobricks would provide individual sections of one large holographic image, heavily reducing the amount of data needed. This could eventually lead to the use of holograms in daily consumer entertainment like movies, games and digital displays.

5. Virtual Reality: After making its dramatic name change, the company once known as Facebook has become Meta. This marks Zuckerberg and his huge team's move into the metaverse – an embodied internet mostly accessed through virtual and augmented reality.

As part of this move, we will start to see Meta putting more time into equipment for accessing this new world – mostly in VR. Announced back in 2021, Meta has been developing a new headset under the title 'Project Cambria'.

Unlike the brand's previous VR ventures like the Oculus Quest 2, this won't be a device for the average consumer, instead looking to offer the best VR experience they can make. The Cambria has been reported to be focused on advanced eye and face tracking (to improve accuracy of avatars and your in-game movements), a higher resolution, increased field-of-view and even trying to make the headset significantly smaller.
Between Meta, Google, Sony and plenty of other big tech companies, VR is getting lots of funding right now and will be seeing drastic improvements in the next couple of years.

Conclusion

The most important factor in survival is "The Will to Survive". To compete and survive in high-tech era, one must learn such technologies so that they can posses efficient skills and adapt to the high-tech future which is ahead of us. Thus, one will be able to survive and compete in the high-tech era.

Aditya Pandey & Punit Kumar

Chapter 3

Adapting With The Technological Swift

Technology can, does and already has influenced how humans have evolved. The world in which we live, work, and play is changing as a result of the fast advancement of technology. Nearly every day, new innovations and paradigms are created. Modern technology is radically changing traditional ideas that the human race has relied on for decades or even centuries.

Computers were barely beginning to appear fifty years ago and were exclusively employed for certain scientific or research tasks, but today they are omnipresent. Around us, computer power may be found in a variety of forms and sizes. Ones that are compact and portable are found in cellphones. Large ones include desktops and mainframes. And recently, with cutting-edge technologies like Google Glass or Snap Spectacles, this computational capability has entered our wearable technology as well. Whether we like it or not, we can certainly see how the technology revolution is changing our daily life.

However, an intriguing query emerges: What long-term repercussions would this transformation have? Not talking about the consequences for a decade or a century from now; rather, talking about what they mean for a human evolutionary timeline that may last for millennia. We see that the rhythms of our lives are altering as a result of technology. We used to be quite physically active, but now days much of our labour is done by technology. The next stage of our species' evolution may be mostly driven by the modifications to our behavioural patterns and adaptations brought on by these technological developments.

People of the Future

The inhabitants of the earth are divided not only by race, nation, religion or ideology, but also, in a sense, by their position in time. Examining the present populations of the globe, we find a tiny group who still live, hunting and food-foraging, as men did millennia ago. Others, the vast majority of mankind, depend not on bear-hunting or berry-picking, but on agriculture. They live, in many respects, as their ancestors did centuries ago. These two groups taken together compose perhaps 70 percent of all living human beings. They are the people of the past.

By contrast, somewhat more than 2.5 percent of the earth's population can be found in the industrialized societies. They lead modern lives. They are products of the first half of the twentieth century, molded by mechanization and mass education, brought up with lingering memories of their own country's agricultural past. They are, in effect, the people of the present. The remaining two or three percent of the world's population, however, are no longer people of either the past or present. For within the main centers of technological and cultural change, in Santa Monica, California and Cambridge, Massachusetts, in New York and London and Tokyo, are millions of men and women who can already be said to be living the way of life of the future. Trend makers often without being aware of it, they live today as millions more will live tomorrow. And while they account for only a few percent of the global population today, they already form an international nation of the future in our midst.

They are the advance agents of man, the earliest citizens of the world-wide super-industrial society now in the throes of birth. What makes them different from the rest of mankind? Certainly, they are richer, better educated, more mobile than the majority of the human race. They also live longer. But what specifically marks the people of the future is the fact that they are already caught up in a new, stepped-up pace of life. They "live faster" than the people around them. Some people are deeply attracted to this highly accelerated pace of life going far out of their way to bring it about and feeling anxious, tense or uncomfortable when the pace slows.

They want desperately to be "where the action is." (Indeed, some hardly care what the action is, so long as it occurs at a suitably rapid clip.) James A. Wilson has found, for example, that the attraction for a fast pace of life is one of the hidden motivating forces behind the much publicized "brain-drain" the mass migration of European scientists to the United States and Canada. After studying 517 English scientists and engineers who migrated, Wilson concluded that it was not higher salaries or better research facilities alone, but also the quicker tempo that lured them. The migrants, he writes, "are not put off by what they indicate as the 'faster pace' of North America; if anything, they appear to prefer this pace to others."

Similarly, a white veteran of the civil rights movement in Mississippi reports: "People who are used to a speeded up urban life can't take it for long in the rural South. That's why people are always driving somewhere for no particular reason. Traveling is the drug of The Movement." Seemingly aimless, this driving about is a compensation mechanism.

Understanding the powerful attraction that a certain pace of life can exert on the individual helps explain much otherwise inexplicable or "aimless" behavior. But if some people thrive on the new, rapid pace, others are fiercely repelled by it and go to extreme lengths to "get off the merry-go-round," as they put it. To engage at all with the emergent super-industrial society means to engage with a faster moving world than ever before. They prefer to disengage, to idle at their own speed. It is not by chance that a musical entitled stop the world - I want to Get off was a smash hit in London and New York a few seasons ago.

The quietism and search for new ways to "opt out" or "cop out" that characterizes certain (though not all) hippies may be less motivated by their loudly expressed aversion for the values of a technological civilization than by an unconscious effort to escape from a pace of life that many find intolerable. It is no coincidence that they describe society as a "rat race" a term that refers quite specifically to pacing.

Older people are even more likely to react strongly against any further acceleration of change. There is a solid mathematical basis for the observation that age often correlates with conservatism: time passes more swiftly for the old. When a fifty-year-old father tells his fifteen-year-old son that he will have to wait two years before he can have a car of his own, that interval of 730 days represents a mere 4 percent of the father's lifetime to date. It represents over 13 percent of the boy's lifetime. It is hardly strange that to the boy the delay seems three or four times longer than to the father.

Similarly, two hours in the life of a four-year-old may be the felt equivalent of twelve hours in the life of her twenty-four-year-old mother. Asking the child to wait two hours for a piece of candy may be the equivalent of asking the mother to wait fourteen hours for a cup of coffee.

There may be a biological basis as well, for such differences in subjective response to time. "With advancing age," writes psychologist John Cohen of the University of Manchester, "the calendar years seem progressively to shrink. In retrospect every year seems shorter than the year just completed, possibly as a result of the gradual slowing down of metabolic processes." In relation to the slowdown of their own biological rhythms, the world would appear to be moving faster to older people, even if it were not.

Whatever the reasons, any acceleration of change that has the effect of crowding more situations into the experiential channel in a given interval is magnified in the perception of the older person. As the rate of change in society speeds up, more and more older people feel the difference keenly. They, too, become dropouts, withdrawing into a private environment, cutting off as many contacts as possible with the fast-moving outside world, and, finally, vegetating until death. We may never solve the psychological problems of the aged until we find the means through biochemistry or re-education to alter their time sense, or to provide structured enclaves for them in which the pace of life is controlled, and even, perhaps, regulated according to a "sliding scale" calendar that reflects their own subjective perception of time.

Much otherwise incomprehensible conflict between generations, between parents and children, between husbands and wives can be traced to differential responses to the acceleration of the pace of life. The same is true of clashes between cultures.

Each culture has its own characteristic pace. F. M. Esfandiary, the Iranian novelist and essayist, tells of a collision between two different pacing systems when German engineers in the pre-World War II period were helping to construct a railroad in his country. Iranians and Middle Easterners generally take a far more relaxed attitude toward time than Americans or Western Europeans.

When Iranian work crews consistently showed up for work ten minutes late, the Germans, themselves super-punctual and always in a hurry, fired them in droves. Iranian engineers had a difficult time persuading them that by Middle Eastern standards the workers were being heroically punctual, and that if the firings continued there would soon be no one left to do the work but women and children.

This indifference to time can be maddening to those who are fast-paced and clock conscious. Thus Italians from Milan or Turin, the industrial cities of the North, look down upon the relatively slow-paced Sicilians, whose lives are still geared to the slower rhythms of agriculture. Swedes from Stockholm or Göteborg feel the same way about Laplanders.

Americans speak with derision of Mexicans for whom manana is soon enough. In the United States itself, Northerners regard Southerners as slow-moving, and middle-class Negroes condemn working-class Negroes just up from the South for operating on "C.P.T." Colored People's Time. In contrast, by comparison with almost anyone else, white Americans and Canadians are regarded as hustling, fast-moving go-getters.

Populations sometimes actively resist a change of pace. This explains the pathological antagonism toward what many regard as the "Americanization" of Europe. The new technology on which super-industrialism is based, much of it blue-printed in American research laboratories, brings with it an inevitable acceleration of change in society and a concomitant speed-up of the pace of individual life as well. While anti-American orators single out computers or Coca-Cola for their barbs, their real objection may well be to the invasion of Europe by an alien time sense. America, as the spearhead of super-industrialism, represents a new, quicker, and very much unwanted tempo.

Precisely this issue is symbolized by the angry outcry that has greeted the recent introduction of American-style drugstores in Paris. To many Frenchmen, their existence is infuriating evidence of a sinister "cultural imperialism" on the part of the

United States. It is hard for Americans to understand so passionate a response to a perfectly innocent soda fountain. What explains it is the fact that at Le Drugstore the thirsty Frenchman gulps a hasty milkshake instead of lingering for an hour or two over an aperitif at an outdoor bistro. It is worth noticing that, as the new technology has spread in recent years, some 30,000 bistros have padlocked their doors for good, victims, in the words of Time magazine, of a "short order culture." (Indeed, it may well be that the widespread European dislike for Time, itself,is not entirely political, but stems unconsciously from the connotation of its title. Time with its brevity and breathless style, exports more than the American Way of Life. It embodies and exports the American Pace of Life.)

HOW EVOLUTION WORKS

According to Charles Darwin, evolution is the change in a species' features over a number of generations, which might occur over a long period or within a few years. Natural selection plays an important role in this process of development. The premise of Charles' theory of evolution is that all species undergo progressive change throughout time. Genetic variety in a population that influences an organism's physical features is necessary for the intricate process of evolution. "Natural selection" refers to the process through which succeeding generations repeatedly acquire favorable traits.

A person is more likely to live, get food, escape predators, and prevent sickness if they have the traits that are most suited to their environment and the evolutionary lottery. They are more likely to procreate and convey their genes to their offspring than other members of the species.

However, the losers in this race who are ill suited to their environment are less likely to live and procreate. As a result, there is a lower likelihood that their genes will be passed on to the following generation. As a consequence, the members of a certain species that are most adapted to their environment as individuals survive, and given enough time, the species will eventually develop. Darwin referred to this phenomenon as "survival of the fittest." Even if Darwin's theory was able to explain the evolutionary cycles of many species, including homo sapiens, it appears that it is no longer working, especially when it comes to humans.

Furthermore, this problem can be directly linked to the expansion of technology's influence in the healthcare industry. By using cutting-edge treatments like antibiotics, some dangerous illnesses have been totally wiped off. Due to this, even the most frail and defenceless individuals among us may live rather comfortable lives. They occasionally also have the ability to procreate via cutting-edge methods like in vitro fertilization (IVF). As a result, this situation no longer involves the survival of the fittest.

A large majority of people may access and obtain the majority of customary medical treatments, medications, and surgical operations (except for an unfortunate segment of the global population who still do not have access to affordable and reliable healthcare). However, let's pretend for the moment that every human being has access to a safe, healthy environment. Even with these common benefits, we can still tell that every person is unique in some way. While some of us are quicker to adopt new technologies and trust them, others take longer to do so.

These characteristics set one individual apart from others in the modern world in terms of efficiency. These adaptive qualities may have an impact on a person's quality of life as well as other aspects of their life, such as their financial situation and position in modern society. This means that everyone, regardless of physical limitations, who has a strong mind may flourish in this world.

TECHNOLOGY: A BOON

No longer do we need to go hunting for food. We no longer have to contend with powerful predators every day. Our bodies are now subjected to far less physical demands and pressures. In the workplace, in agriculture, and in transportation, machines now handle the heavy work. In the modern society, mental aptitudes are what set people apart. The intellect is the focus of attention while physical capability is given less weight.

The automated machines are created and controlled by the human intellect. Information is processed by our thoughts at a rate that is constantly accelerating. We can draw on this information over the centuries and transmit it orally or in writing from one generation to the next. As a result, the human species is rapidly adding to its body of knowledge. Therefore, it wouldn't be inaccurate to suggest that the human mind is gradually changing to adapt to this world of mechanization, complexity, and knowledge.

The process of evolution itself is a gradual one. However, by closely monitoring this occurrence across a number of species over extended periods of time, we can forecast how it will most likely develop. The human mind is the organ that is employed the most frequently in the modern world, as we established above. From the difficult calculations we must perform at work to the quick decisions we must make every day. Accordingly, it is likely that these two elements will influence how our human brain evolves.

The human brain may eventually begin to prioritise its analytical abilities over its capacity for information retention (because this capability is already provided to us by computers). It's like stating that although people will have stronger CPUs and RAM, their hard drives will be comparatively smaller.

Other organs and limbs may also see a significant impact. For instance, our hands, legs, and feet could experience significant alteration. In order to build a home and go on food hunts, our ancestors utilised their hands for physical work. We moved from one continent to another following our food using our legs and feet. However, these mobile organs are currently employed for far less laborious tasks. Therefore, it is quite likely that adjustments will be made to these organs to increase their effectiveness.

The human species did not suddenly arise. These beings are the results of millions of years of evolution. The many mutations that took place both before and after the 200,000-year-old advent of modern humans to produce the human genome as it is now.

Beyond what is physiologically feasible, there are several additional technologies that are improving our physical and mental capacities. Examples of this include Google Glass and bionic body components. One's mind received cognitive knowledge via Google Glass that was not humanly conceivable to acquire in such a short period of time in real life. On the other hand, a person with bionic limbs and legs can withstand weights that are beyond the capabilities of regular people.

There are plethora of examples that shows the need of updating ourselves with the ongoing technological advancement. The sudden covid outbreak brought in multiple up-gradations in high-tech world. No one would have wondered that physical classes could be replaced by an online one, people would attend offices at home, and students having their convocation virtually.

But this all happened, people adapted to the situation and this ideally should be the whole need of the hour. Starting from Australopithecus (hominid) to now as a civilized homo sapiens, a man has nothing but evolved according to the need and this should be continued in future too, for if a man will not evolve and update himself with the ongoing technological revolution, soon he will return to the scratch/shunya.

Conclusion

In conclusion, it is safe to assume that the fast rise of technical breakthroughs will undoubtedly have an impact on human evolutionary cycles. Because of the profound changes that these technologies have already made to life's patterns, the rate of evolution as a whole has accelerated. Because healthy, productive lifestyles are attainable for people of all physical capacities, their mental genes can change and produce superior outcomes. Artificial evolution is also not too distant in the future, and eventually humans could be able to choose the many features of their kids, such as cerebral aptitude, physicality, and energy levels.

To survive, to avert what we have termed future shock, the individual must become infinitely more adaptable and capable than ever before. He must search out totally new ways to anchor himself, for all the old roots religion, nation, community, family, or profession are now shaking under the hurricane impact of the accelerative thrust. Before he can do so,however, he must understand in greater detail how the effects of acceleration penetrate his personal life, creep into his behavior and alter the quality of existence. He must, in other words, understand transience.

Aradhya Pandey

Chapter 4

Coping With Technology To Survive the Future

"*It is not the strongest of the species that survives, nor the most intelligent that survives. It is the one that is the most adaptable to change.*" - Charles Darwin

Digital technologies are progressing rapidly. In fact, humanity has been making gigantic steps in this sense in the past few decades. Logically, there would be new emerging technologies appearing every once in a while, as well as a few older ones evolving simultaneously (McMinn, 2019).

Digital Technology and Digital Literacy

Teaching literacy means the teaching of the ability to read and write. From these origins, in today's world of new technology and communication, arises the concept of digital literacy. This is a process that allows us to acquire the necessary skills to be competent in new technologies. Therefore, digital literacy concerns the ability to perform different tasks in a digital environmental.

Digital literacy encompasses different skills. For example, locating, analyzing, organizing, understanding and evaluating information; all carried out via digital technology (The Importance of Digital Literacy in Today's World - Website).

McMinn (2019) highlighted few new digital technologies as follows:

1.Virtual Reality (VR) and Augmented Reality (AR)

VR technology provides the user with a complete immersion in the virtual world. It shuts off the real world and gives a complete experience. By using external devices that usually come in the form of a headset, the user can "teleport" themselves to a completely or partially different reality. AR technology adds various digital elements to a live view. This is usually done with the help of a smartphone's camera.

MR technology combines VR and AR to create an experience where real-world and digital objects interact. It is the newest of the three and is now being developed. Microsoft's HoloLens is an example of MR technology.

2. Voice Assistant:

Voice Assistant , also called virtual assistant, digital assistant, and AI assistant is a program that understands natural language voice commands and completes the tasks the user orders. Some popular examples of such applications include Apple's Siri, Google's Assistant, Amazon's Alexa, Samsung's Bixby, and Microsoft's Cortana.

Voice assistants are now mostly present in smartphones allowing owners to lead an easier life by completing tasks usually performed by secretaries. For example, a voice assistant can set up your alarm, remind you about a task, write down the text you are saying, and so on.

3. Motion UI:

In comparison to static design, motion design or motion user interface (aka motion User Experience) is a lot more engaging. In fact, motion design can help in many different aspects of interacting with users on your website: such as Welcoming users, Information about the actions; confirming the activities; adding fun Elements; Refreshing content; and Feeding loop.

4. Artificial Intelligence (AI), Machine Learning (ML) and Deep learning (DL)

All three are quickly becoming an integral part of the business world, and though they may seem similar, these technologies are actually different from each other.

Artificial Intelligence (AI) applies human intelligence to machines. AI encompasses everything from symbolic artificial intelligence to deep learning. To put it simply, whenever a machine completes a task based on a set of rules that solve problems (aka algorithms), it is called AI.

Machine Learning (ML) empowers computer systems with the ability to learn. It focuses on teaching computers to learn by themselves with the help of big data and then make accurate predictions.

Deep Learning (DL) is a subset or the next evolution stage of ML. DL algorithms were inspired by the information processing patterns in our own brains. For example, the way we make decisions can be reflected in a DL algorithm.

5. Blockchain

Just like some other new digital technologies on this list, blockchain has been around for a while (a few years at maximum). To put it simply, it's a technology that allows both businesses and individuals to make instant transactions without any middlemen such as bank.

The biggest advantage is that blockchain transactions are completely secure and can be kept as a record of what happened. The codes used in this technology ensure that the record cannot be altered after the transaction is complete.

Blockchain technology is still being tested and developed. It hasn't been universally accepted either, which means that we are yet to experience its full potential. Most people are still figuring out how they could benefit from it. This is one of the most promising new digital technologies.

6. Internet of Things (IoT)

The Internet of Things (IoT) is focused on connecting everything in the world to the Internet. That definition is as simple as it can get as you will mostly encounter complicated technical explanations. As defined by IoT for All, "IoT is a system of interrelated computing devices, mechanical and digital machines, objects, animals or people that are provided with unique identifiers and the ability to transfer data over a network without requiring human-to-human or human-to-computer interaction." Once something is connected to the Internet, it can now send and receive information in order to perform various tasks or access certain data. For example, you can listen to a song on your smartphone after searching for it on the Internet.

7. Progressive Web Apps (PWAs)

Progressive Web Apps (PWA) are, in fact, websites that look and feel like they are apps. Instead of users downloading a mobile app to access all the information and capabilities, they can now simply visit the website.

PWAs use progressive technologies that allow you to create websites that will function almost the same way as mobile apps. PWAs are said to increase conversions and page visits while keeping visitors on your website longer. Simply put, PWAs are the perfect option for those who don't have the budget to develop a separate app for their business.

8. Fifth Generation Technology (5G)

Fifth Generation Technology (5G) is the fifth generation of mobile networks. It is a significant evolution of the 4G LTE networks used today. 5G is currently being designed to meet the constantly growing data and connectivity in modern society

5G is predicted to operate along with the current 4G networks at first. After that, it will evolve into fully independent networks. 5G has faster connectivity, greater capacity, and fast response time or latency. In order to receive and transmit voice and data, 5G uses radio waves or radio frequency also known as RF (McMinn, 2019).

These digital technologies such as Connectivity, Data Analytics, Artificial Intelligence (AI) and the Internet of Things (IOTs) are rapidly changing our world and environment, and according to Moore's Law, the pace of change will only grow (Gonda, 2018). A 2017 IBM report noted that 90% of the world's data had been generated in the past two years. Just like we couldn't have envisioned 2018 in the year 2000, we can't quite know what our habitat will be like in 2025 or 2035 (Gonda, 2018).

Stowers (2022) highlighted that in the future, machines are expected to handle more process-oriented tasks, leaving humans with higher-level tasks like analyzing, problem solving and interacting with other people. A study by PwC says that 38% of jobs are at risk of being automated in some capacity by 2030.

This means that many people in the near future may be working alongside artificial intelligence (AI). But to be able to perform well in this new business world, employees must focus on digital upskilling and reskilling to ensure continuous success (Stowers, 2022).

Digital Upskilling and Reskilling and Survive the Future

From Harper Collins Dictionary: Life-Long Learning (LLL) is the provision or use of both formal and informal learning opportunities throughout people's lives in order to foster the continuous development and improvement of the knowledge and skills needed for employment and personal fulfillment (Laal, 2011). Nowadays, there is an increasingly important basic skill in ever-changing technological universe: ability to learn and adapt to the needed new skills and training (OECD, 2007).

In the 21st century, skills have become more important than knowledge and information of particular subjects. Life skills, career skills, innovative and project-oriented academic works have become primarily important. Moreover, the learners should be competent enough in skills such as using technology for learning purpose and gaining adequate knowledge of different resources. Learners lacking proficiency in these areas probably may not be able to cope with the 21st century demands and job opportunities being offered by national and Multinational Corporations (MNCs). In this context, an American writer and futurist, Toffler (1970) says, "The illiterate of the 21st Century will not be those who cannot read and write, but those who cannot learn, unlearn and relearn". Similarly, it is also said that computer literacy has occupied a certain position and demand in the contemporary world. People who are computer-savvy are called germane to the contemporary world (Gore, 2013).

Digital literacy provides an individual with core capabilities to achieve valued outputs in life. It is a critical enabler of economic transformation as it promotes employment opportunities through the ability to access digital content and online services. A key point raised by UNESCO, is that digital literacy improves one's employability because it is considered a 'gate' skill required by employers. It is a catalyst for individuals to acquire other valued outcomes (Chetty, et.al, 2018),

Upskilling is defined as helping workers adapt their more traditional skills and capabilities to modern technologies that underpin digital transformation efforts, while reskilling provides the employees with new skills, according to Akhtar (O'Donnell, 2020).

In Fillmore (2021) view, digital upskilling contributes to sustainability because it ensures continuity of workforce for society. The Covid- 19 pandemic threatens to increase the digital skills gap. Companies who initiate upskilling themselves will reap the economic benefits. Fillmore (2021) further highlighted that companies that digitally upskill their employees are a key part of social sustainability.

To conclude that, in general, there are many benefits of new technology, but on the other hand, as studies (Chadha & Saha 2020; Mumtaz and Muthkoor, 2021; Digital Literacy - Wikipedia) have indicated that digital literacy is low highlighting the digital divide in especially in India. Also there are urban - rural; age group wise and gender wise differences are observed in digital literacy. Digital illiteracy may lead to stress condition i.e., techno stress condition.

Techno stress has been defined as the negative psychological link between people and the introduction of new technologies. People experience techno stress when they cannot adapt to or cope with the information technologies in a healthy manner.

They feel compulsive about being connected and sharing constant updates, feel forced to work- related information in real- time, and engage in almost habitual multi- tasking. They feel compelled to work faster because information flows faster and have little time to spend on sustained thinking and creative analysis (Wikipedia Techno stress website).

Thus, to bridge this digital divide there is a need to have digital literacy. There are various levels of digital literacy, range from the most basic or elementary level, for example to knowing how to publish on Instagram; Intermediate level i.e. to be more efficient, for e.g. learning to program on Facebook; and Higher level for e.g. creating your own digital content (Website: The Importance of Digital Literacy in Today's World).

In today's world, where the use of new technologies prevails, it's almost essential to, at the very minimum, know what digital literacy consists of. To know that it's a new way of connecting with people, of communicating both with them and the world. In addition, digital literacy goes beyond learning digital tools and programs and has great benefits.

For example, it:

Develops critical thinking: By accessing more information, you can better contrast it, thereby fostering critical thinking. Allows access to better jobs. Knowing and mastering ICT (Information and Communications Technology) today increases the possibility of accessing more and better-paid jobs.

Favors social inclusion and development. In fact, a lack of access to technologies creates more differences in the cultural, economic, social, and educational spheres.

Improves the day-to-day. New technologies can be integrated into your day-to-day life and improve your quality of life, making it easier and more comfortable (Website: The Importance of Digital Literacy in Today's World (Website: exploringyourmind.com)

To sum up, we are a species whose habitat is now ruled by technology, and it is changing at a rapid pace (Gonda, 2018). As Charles Darwin once said, "It is not the strongest of the species that survives, nor the most intelligent that survives. It is the one that is the most adaptable to change." As humans, when we are forced to accept new social and work dynamics, we can adapt very quickly.

We innovate new industries, create new jobs, and develop new ways of working. We do anything and everything we can to survive and thrive when things are against us. Ultimately, there is no tool more powerful for survival than our ability to adapt (Website: Never Underestimate the Power of Human Adaptability).

Adaptation refers to traits of an organism that increase its fitness to survive and reproduce (Basic Concepts of Ecology-IGNOU Website). As we live in the digital world, technology governs every aspect of life (New Age Technology and Digitization- Website) and evolving rapidly. To sustain and survive in the evolving digital world individual must become infinitely more adaptable to new technology circumstances by digital upskilling and reskilling to increase our fitness.

Dr. Anuradha Hanumant Deshpande

Assistant Professor, SVVV, Indore

Chapter 5

Future of Education In Digital World - Trends & Techniques

The task of building an enlightened, strong and prosperous nation rests on the shoulders of its children who are to be cherished, nurtured and developed with tenderness and care. Education has always played this important role and has thereby emerged as a natural characteristic of human societies. It has contributed to the shaping of the destinies of societies in all the phases of their development and has itself never ceased to develop. It has been the torch bearer of humanity's most noble ideals. In this sense, as an agent for social change, education necessarily reflects its main ethos, aspirations and concerns.

Education aims at making children capable of becoming responsible, productive and useful members of society. Knowledge skills and attitudes are built through learning experiences and opportunities created for learners in school. It is in the classroom that learners can analyse and evaluate their experiences, learn to doubt, to question to investigate and to think independently. Education in India has undergone a wide range of change from Gurukul system to present context where technology plays an integral part of teaching- learning process.

It is relevant to note that the globalization and technological change have created a new global economy "powered by technology, fueled by information and driven by knowledge. The emergence of this new global economy has serious implications for the nature and purpose of educational institutions. As the half-life of information continues to shrink and access to information continues to grow exponentially, schools cannot remain mere venues for the transmission of a prescribed set of information from teacher to student over a fixed period of time. Rather, schools must promote "learning to learn" i.e., the acquisition of knowledge and skills that make possible continuous learning over the lifetime.

Concerns over educational relevance and quality coexist with the imperative of expanding educational opportunities to those who made most vulnerable by globalization developing countries in general; low-income groups, girls and women, and less-skilled workers in particular.

There has been a phenomenal impact of technology on world that is transforming the classrooms of today. Educators are now adopting different teaching methods to fit learning styles, interests, and expectations of the learners & parents. Leveraging technology, educational software, and new learning tools, educators across the world are increasingly blending in-class and online learning strategies.

New Trends in Future Education: According to RMIT School of Education Professor Tricia McLaughlin about four key ways education is set to change in the near future.

1. Connectedness, collaboration and co-creation

The concept of a teacher standing in front of a room full of students who listen and respond to direction is increasingly a thing of the past. While not an entirely new approach, student learning spaces will supersede the typical classroom that we know today. This will see students become partners or co-creators of their own learning.

“Experiences that allow collaboration, communication and teamwork for all students often happen beyond classroom walls. We need to facilitate for these experiences in context, and our classrooms need to be a reflection of this," McLaughlin says. They will be set up to allow collaboration to occur on learning projects between individuals, small groups or larger groups.

Classrooms will coexist as physical spaces and online, flipping the current learning model upside down so that students can learn at home and spend class time collaborating and applying their knowledge to real-life issues.

2. Anywhere, anytime learning

As we ride the wave of the digital era, it is becoming easier to get connected with a global reach. A world of information is at your fingertips with the click of a button or a simple voice command, and, as technology continues to advance, students need to grow their learning with it.

Technology is no longer a motivating factor when it comes to learning – it is a must. It's something that needs to be incorporated in the future of education to ensure students are equipped with the skills to cope in a world dependent on technology.

While some argue that technology in our classrooms creates lazy, disconnected students, McLaughlin believes this is a myth. She says technology has created endless boundaries of where learning can occur, with whom and why.

"The reality is, classrooms can be anywhere anytime. Students can be working on projects in virtual contexts with other students from around the world at any given moment," she says. Technological advances have enabled inter-connectedness of information and people with the touch of a button.

Education in the future will need to demonstrate how technology can be used to students' advantage, as well as teach future generations how to handle problems that arise from it.

"Technology can change learning forever and we need to embrace it and manipulate it to our advantage," McLaughlin says.

3. Customisation for a learner-first approach

Alongside our changing notions of what constitutes a classroom, our ideas about the way teaching is delivered must also be reshaped. "Most professions treat each individual's case differently – each patient of a doctor has individualised treatment plans. Education should be no different," McLaughlin says.

She says the old 'one model of teaching and learning fits all' is outdated and has no place in the agenda for future education. Teachers will become facilitators of learning and students will have more control of their own learning journey.

"In the past, all children did the same work regardless of ability or skills. We now know that this contributes to disengagement, misbehavior and poor outcomes," she says.

As a result, teachers will have individualized learning plans for students, which will enable each student to learn at a pace that best suits their abilities and to engage with content that is most beneficial to them.

A combination of evidence-gathering and feedback from parents, students and other professions will enable these plans to be successfully integrated into the education system. To maximize the potential for individual progress, some elements of teacher-led learning will remain, which will augment traditional learning practices when combined with online digital media.

4. Putting testing to the test

McLaughlin poses the question “what are we testing for?”. Students today are heavily focused on the end result – achieving that high score, receiving a distinction in class, acing those tests. Education of the future will prove what you have been told many times before: results do not define you.

McLaughlin believes that testing on its own can be a dangerous approach. “We too easily jump onto the test results in isolation of what we need to achieve. This can lead us off in the wrong direct to what’s really important,” she says.

Grading is a waste of time if its purpose is solely to point out who is at the top and who is at the bottom. Assessments in the future will be evidence based, using measures that allow learning plans to be drawn up and personalised.

“The big question is ‘why are we grading and for whom?’,” McLaughlin says. “Surely, we want a society in which everyone is able to do the job and elements of it competently. Do exams really prepare us for this future?”

Need for a broad set of knowledge, skills, attitudes and values in action

Students who are best prepared for the future are change agents. They can have a positive impact on their surroundings, influence the future, understand others' intentions, actions and feelings, and anticipate the short and long-term consequences of what they do. The concept of competency implies more than just the acquisition of knowledge and skills; it involves the mobilisation of knowledge, skills, attitudes and values to meet complex demands. Future-ready students will need both broad and specialised knowledge. Disciplinary knowledge will continue to be important, as the raw material from which new knowledge is developed, together with the capacity to think across the boundaries of disciplines and "connect the dots".

Epistemic knowledge, or knowledge about the disciplines, such as knowing how to think like a mathematician, historian or scientist, will also be significant, enabling students to extend their disciplinary knowledge. Procedural knowledge is acquired by understanding how something is done or made – the series of steps or actions taken to accomplish a goal. Some procedural knowledge is domain-specific, some transferable across domains.

It typically develops through practical problem-solving, such as through design thinking and systems thinking. Students will need to apply their knowledge in unknown and evolving circumstances. For this, they will need a broad range of skills, including cognitive and meta-cognitive skills (e.g. critical thinking, creative thinking, learning to learn and self-regulation); social and emotional skills (e.g. empathy, self-efficacy and collaboration); and practical and physical skills (e.g. using new information and communication technology devices).

The use of this broader range of knowledge and skills will be mediated by attitudes and values (e.g. motivation, trust, respect for diversity and virtue). The attitudes and values can be observed at personal, local, societal and global levels. While human life is enriched by the diversity of values and attitudes arising from different cultural perspectives and personality traits, there are some human values (e.g. respect for life and human dignity, and respect for the environment, to name two) that cannot be compromised.

Use of Innovative techniques in Education:

"Pedagogical innovation is significant positive change in enhancing and enriching educational experiences."Pedagogical Innovations are the creation of innovative teachers who believe in high learning outcomes based on cognitive domains. These innovations are occurred where student learning is considered as top most priority. Pedagogical Innovations encourage vibrant approaches to teaching as well as learning in the course of creativity and innovation.

Pedagogical innovations are concerned with innovations in pedagogical decisions, pedagogical practices and pedagogical knowledge. Innovations are basically associated with pure science and technology. But in pedagogical context, it has many forms and is also called scholastic Innovations. Technology may support pedagogical innovations in a better way. Pedagogical practices always require being novelty.

Welder (2014) revealed that unlike technological innovations, these innovations are pedagogical since they composed by pedagogical thinking particularly, and in human relations at the personal motivation of the dedicated professor. Teaching in novel ways that help students to learn and achieve learning outcomes like using emerging technology is also inhabit in the meaning of Pedagogical Innovations.

The modern trends in form of Online courses, e learning technologies, social networking tools, and other emerging technologies are being popular among the young generation and further leads to technology supported pedagogical innovation with full access of information communication technology.

There is variety of digital tools and technologies available for making pedagogical designs and planning instructional practice for excellent educational experiences. Here we will discuss some digital tools which can be of use in this regard:

Virtual learning environment Tools:

a. **Learning Management system:** When learning material and learning experiences are transacted through internet or web, it is called Virtual Learning Environment. This is also called Learning Management System and Content Learning Management System. This system consists of learning resources, assessment, evaluation, student tracking and discussion form.

These virtual learning environment tools can be accessed on campus as well as distant places. These are helpful to support students in leaning at any pace and place. From this platform, institutions can teach their students not only with the traditional form but through flipped classroom approach also.

Learning Management System as a Virtual Learning Environment Tool is a software application which is used for delivering and managing e content information of video, text etc. The main characteristics of this LMS are Rubrics, teacher and student role, a discussion board, educational resources along with variety of functionality. It is of both types; paid as well as free of cost. Example:

- ✓ Model
- ✓ Course builder
- ✓ Google Classroom
- ✓ Ed mod
- ✓ Blackboard

b. **Lecture Capture:** Lecture capture is quite a broad term. It is concerned with a process which is used for recording classroom lectures as a video lesson. After the class, these videos can be seen and reviewed by the students. It consists of a variety of solutions, applications, software and hardware.

Lecture capture is an excellent tool of Virtual Learning Environment for moving lecture outside from the class for flipped classroom learning. Lecture Capture Tools allow instructors to create recordings of classroom lectures or presentations, including audio, video and screen content, and share them with students online and in Blackboard.

- ✓ Screencastify
- ✓ Screencast-o-Matic
- ✓ Cam Studio
- ✓ Webineria
- ✓ Panopato
- ✓ Presenataion Tube

c. **Webinars:** Webinar consists of web and seminar. A webinar is an event which is conduced or organized through internet and attended by online viewers. This term is similar to web event, online seminar, web lecture but to some extent different from webcast. The basic feature of this tool is that through webinar, online presentation can be changed into a real time conversation from anywhere in the world and via its interactive element, online participants may give, receive and discuss information in real time.

A huge number of participants can participate in webinar and able to share documents and slides etc. In short it is called Web based seminar from which slide presentation, lecture are conducted through the internet along with video conferencing tools /software. It is highly useful when any institution hosts a lecture through webinar and while presenting and participating in this, participants or attendees share slides and documents with everyone. In hosting webinar, there may be use of live streaming options and we can also record and publish to you tube or any LMS. Example of this is:

- ✓ GoTo webinar
- ✓ Anymeeting
- ✓ Zoom

d. **Interactive Web Conferencing:** Like Webinars, Interactive Web Conferencing tool is also an online collaboration tool. It consists of an audio- video web conferencing software or application that combines all the amenities through internet. This is less collaborative as compared to webinar but highly interactive in nature. Web conferencing is made possible by Internet technologies, particularly on TCP/IP connection. The main feature of the web conferencing is that it works best for small to medium group collaboration.

It is used for conducting and organizing conference or meeting for distantly present members to share online collaboration of different features including documents, data, slides, report and video etc. It reduces the time and cost of travel for different meetings or conferences. There is web collaboration service features in web conferencing to record Interactive Web Conferences, notes, participants Questions &Answers, chats, polling, and video conferencing with a web cam.

A web conference is used for an event where multiple computer users can communicate with each other at the same time using cameras over the internet. Web conferencing makes the most sense for small and collaborative meetings (usually under 50) by involving a good amount of participation. A webcast is a live or on demand presentation that is streamed over the internet. It is ideal for targeting very large audiences (unlimited size). It is offers high quality, live session and makes use of interactive features like survey and polls, and Question &Answers.

- ✓ Bigbluebutton
- ✓ Google Hangouts
- ✓ Blue Jeans
- ✓ WorkplacebyFacebook
- ✓ SkypeforBusiness
- ✓ GSuite

e. **Electronic books:** Electronic Book is also synonymous of E-book. The main features of the E Book is that it is in digital form and exhibits the characteristics of text, images and videos in e content form. This may be readable on computers and electronic device. Sites with free eBooks as follows:

- ✓ Project Gutenberg
- ✓ Google Books
- ✓ Free TechBooks
- ✓ O'Reilly Open Books
- ✓ Bookboon
- ✓ Free Computer Science Books
- ✓ Wikibooks
- ✓ FreeBooks Doctors
- ✓ DevFree Books
- ✓ Free-Ebooks.net
- ✓ Internet Archive

f. **Screen sharing:** As a supporting tool of Virtual Learning Environment, this is used for sharing slides, presentations and documents for easy collaboration in meetings and training. Screen sharing tools make collaboration easy, visual and interactive as it would be appeared in person.

Through this, design, document, or website can be presented exactly as it displays on one's screen. Screen sharing software allows people from a variety of distributed locations to view another person's computer screen in real-time conference:

- ✓ GlobalMeet
- ✓ SkyPe
- ✓ Zoom
- ✓ Redbbooth
- ✓ Teamviewer

g. **File sharing and document management systems:** In virtual and collaborative environments, there should be provisions of file sharing and document management system like Dropbox and G Drive. A document management system is a computerised and robotic method to organize, secure, capture, digitize, tag and complete tasks and assignments with files. The most beneficial characteristics of document management system is just like a cloud storage. Example of these are:

- ✓ E File Cabinet
- ✓ Dropbox
- ✓ Google Drive
- ✓ Micro Soft One drive
- ✓ File cloud
- ✓ work drive

h. **Social Media Tools:** The statement about social media is that Social Media Technology has empowered but also accelerated social sharing and content creation. Social media tools help in boosting the educational experiences in more sophisticated way because in 21st century our pedagogical concern and skills must be ready to grapple with the significant changes associated with technology supported pedagogical innovations.

In this new age, where learners are digital native, social media may be highly utilized as a pedagogical tool where students would not get only information from his teachers but share, collaborate, reflect and apply with broader community and learns life skills. Teacher is only a facilitator to students and all desired information which they require, they would get it from Google easily. Social media helps to communicate and cultivate relationships among society for schools and institutions.

When educational experiences are told in a digital story method on social media platform, it creates magical rhythm in a vibrating and dynamic manner. There are various tools from which teachers can sharpen 4Cs of 21st century skills as Critical Understanding, Communication, Creativity, Collaboration etc. Example of these are as follows:

- ✓ Facebook
- ✓ Twitter
- ✓ Bloogger
- ✓ Youtube
- ✓ Instagram
- ✓ Vimeo
- ✓ TeacherTube
- ✓ Pinterest
- ✓ Voice thread
- ✓ Podcasts
- ✓ EduBlogs

I. **Professional Development Platforms:** It is stated that continual and perpetual education is the key to success for further professional development. Professional Development platforms are essential and provide opportunities to explore new knowledge, latest trends and contemporary issues that connect with professional learning communities through technologies.

Teachers may get best knowledge through Professional Development platforms and can become popular by having specific training, recognised courses and advanced professional skills to improve and enhance professional knowledge, competence, skill and effectiveness.

- ✓ Research gate
- ✓ Academia education
- ✓ Wiki
- ✓ Wikispace

j **Digital Resources to support teaching and learning:** Digital resources can be highly utilized in a variety of ways to support teaching and learning. Electronic grade books, digital portfolios, learning games, and real-time feedback on teacher and student performance can be used to empower learning.

Digital Technology or E Learning technology has given birth to various tools that teachers may use within the classroom and outside the classroom to enhance student learning experiences. Here are some examples of digital resources to support teaching and learning.

a. Presentation Programme: This is a software package which is used to exhibit or display information in the form of a slideshow or video. Basically, It has three main characteristics as text editor, graphic images, music tune that support teaching learning resources in a magnificent way. Example of these are as follows: PowerPoint, PREZZI, Slide share, Keynote, Powtoon, Adobe spark.

b. Using YouTube Video in Teaching and Learning: By using video for educational experiences, research revealed that student preserve maximum information, understand the concepts easily. By utilization of

YouTube video as educational resources, we can explore new connections between curriculum topics and world outside from the classroom.

c. Storyboard: A storyboard is a graphic organizer in the form of illustrations or images displayed in sequence for the purpose of pre-visualizing of motion picture, animation, motion graphic or interactive media sequence. (Wikipedia) It is also called a pictorial framework for creating video. It consists of a series of thumbnail images that convey whole story from starting to last. There also exits written script about what's going on in each frame. When it finishes, the storyboard looks like a comic strip. Storyboards can be simple or complex. It is hand-drawn, but we can use story board software to create images for making video.

Advantages of digital technology in education

Digital technology is transforming today's classroom very quickly in many aspects. Some schools have already incorporated digital technology in teaching and assessment. Various research studies report many benefits of using digital technology. Some of them are listed below:

1. Improvement in the retention of the content: The use of various devices and aids has made learning simpler, easier and the ability to retain content and information of the student's has shown considerable improvement.

2. Save Time and Energy: Digital presentations and animations save teacher's time in the classroom. In addition, teacher can use his teaching learning material in the form of videos and e-content with adequate updates over and over again. Thus, the time and energy of the teacher can be saved.

3. Engagement and in-depth learning: The student remains more active, interested and engaged in the learning process. Different demonstrations and activities can be shown with the help of digital tools for better understanding.

4. Easy Access: The best thing about the digital technology is that learners can easily access information from anywhere beyond geographical constraints. This provides flexibility in the learning, as the student can use and retrieve the content in accordance with his capability and interest. Students can even participate in video conferencing with teachers and tutors in real time.

5. Open and distance learning: Learning with digital technology provides the best opportunity for the learner to gain education through open and distance mode without compromising the quality.

6. Personalised and Independent learning: Learning theories are based on the fact that every individual is different from one another in their learning style, grasping power and pace of learning. Teacher gets very few opportunities for personalized learning in regular classroom but digital tools can solve this problem very easily.

Artificial intelligence technology allows a teacher to design individualized learning paths for different individuals where they can learn at their own pace. This feature of technology can become landmark for the education of children with special needs. Along with personalized learning this also promotes independent learning which ultimately brings confidence.

7. Economical: Although, technology is expensive at initial stage but it provides many ways to save money. Various organisations and institutions can very easily share and access

their content. Somebody sitting in India can easily access various resources available online around the world.

8. Evaluation and Feedback: Using digital tools, teachers can easily assess the progress of students and can immediately give feedback to students.

9. Interaction and collaboration: In traditional classrooms, interaction between the student and teacher was limited only to the classroom and students' collaboration in any group project was also limited. Digital technology has provided opportunity for both teacher and the student to interact with each other anytime, anywhere.

10. Critical and Independent Thinking: Pupils' ideas may go beyond curriculum, as pupil gets many opportunities to learn from different sources in different forms. Digital learning foster and inspires independent and critical thinking and encourages students to learn further.

11. Automation of Many Tasks: The technology automated the implementation of many times consuming and tedious tasks such as attendance, evaluation and grading etc. This saves teacher's time and effort from being wasted that can be utilized for other tasks.

Digital technology has paved a new future of education accessing and providing more opportunities to all the stakeholders involved in the field of education. The future weaves education fraternity into panorama of skilled, creative and competent individuals. Curriculum teaching and learning already extends well beyond the classroom and will continue to do so, and as education changes to suit the future's needs, the role of a teacher must also adapt and grow. It is each teacher's responsibility to empower students to take risks, be innovative and seize any opportunity thrown their way.

In light of a shift towards a more personalised learner experience, teachers of the future must be prepared to be data collectors, as well as analysts, planners, collaborators, curriculum experts, synthesizers, problem-solvers and researchers.

The two major themes - what we teach and how we teach are likely to be transformed over the next few years into new trends of education. To craft a cohesive strategy, action needs to be taken on multiple fronts. In the immediate term, there must be a mechanism to thoroughly map the ed-tech landscape, especially their scale, reach, and impact. The focus should be on access, equity, infrastructure, governance, and quality-related outcomes and challenges for teachers and students.

The journey from a holistic strategy to its successful application will, no doubt, be a long one. It requires careful planning, sustained implementation, and calculated course corrections.

Dr Madhulika Singh
Assisstant Professor
G.I.P.T.S. Meerut

Chapter 6

Digital Education In Rural India - Need Of The Hour

In the global perspective, there is the rising demand for digital education and digital learning. In India, the same has been realized in the name of 'Digital India' which is a flagship programme of Government of India with the vision of transforming India into a digitally empowered society and knowledge economy. Accordingly, there has been ceaseless march for accelerating the pace of digital education and remote learning initiatives across India in general and rural India in particular.

Various national initiatives are in practice for strengthening the IT-infrastructure, enhancing digitalized solutions and internet connectivity. All these interventions are the urgent needs of both rural and urban India. The services like e-governance, banking, finance, education, healthcare and community development, all are influenced by this digital aspect. Mobile Apps, DTH Recharge, e-ticketing, online shopping are the popular example of digital practices. E-health, e-education, virtual learning and digital information are also the dimensions that undergo digital practices.

The National Digital Health Mission (NDHM) is also one of the landmark initiatives by Government of India based on digital interventions. It facilitates user's health account diagnosis of diseases, tests, consultations, medication and treatment. Thus, Indian economy is the digital economy including digitalized agriculture. As such, digitalization has transformed the agro-industry by digitalizing farmers, their productivity and sustainability.

Credit goes to the Ministry of Education, Government of India for encouraging and implementing diverse remote learning initiatives through innovative use of digital technology and virtual learning. The Central Government along with State/UT Governments has been constantly working towards socio-economic and educational reforms through diverse programmes of digital education and virtual learning.

DIGITAL EDUCATION

Digital technologies are the innovations leading to changes and transformations. When it is applied to education and learning it is called digital education and digital learning respectively. Contemporary COVID-19 pandemic has multiplied the effects of digitization to a greater extent. Generally, digital education is considered as a type of education and learning that is "accompanied by technology or by instructional practice that makes effective use of technology. It encompasses the application of a wide spectrum of practices including blended and virtual learning. Thus, it is the combination of adaptive learning, blended learning, online learning, e-learning, ICT integrated learning, computer aided learning, personalized learning, learning analytics and even use of mobile technologies, iPhone, iPad and other digital devices which have accelerated the pace of digital education and digital learning.

DIGITAL LITERACY

Digital literacy is a pre-requisite for Digital India. It is the key element for e-governance and digital learning and marketing. Accordingly, efforts have been made to ensure digital literacy among Indian citizens.

RELATED TERMINOLOGIES OF DIGITAL EDUCATION

Online Learning: Simply online learning is the learning by accessing available online resources. It might involve the use of the internet and use of online facilities. Online learning can also be done through programs or apps installed on the personal devices of learners or participants.

Web-based Learning: Web-based learning refers to the process and practice of learning by using web browsers.

E-Learning: E-learning is the process of using electronic technologies for teaching-learning processes in which the learning activities take place either entirely or partially online.

Blended Learning: Blended learning generally combines virtual learning with traditional classroom learning. It is an approach 'that uses multiple methods to deliver learning by combining face-to-face interactions with online activities.

Distance Learning: It is the process of learning from a distance in which the participants are physically separated. In other words, distance education provides instruction to a person/persons in rural and remote areas who is learning in a place and at a time different from that of the teachers and the other learners.

Virtual Learning: Virtual learning and virtual classrooms are the unique innovative learning experience for people in rural areas that facilitate learning by using 'computers and/or the internet both outside and inside the facilities of the educational organization. Thus, virtual learning postulates that the quality of online education depends on the proper use of digital technologies and digital contents.

CHALLENGES OF DIGITAL EDUCATION IN RURAL INDIA

Transforming India, especially RURAL INDIA through digital interventions is not free from challenges. It will have a transformational effect on Indian society and Indian economy. The major challenges are technical, organisational and economic.

- ***Technical Challenges:*** Use of technology, application of technology and even availability of appropriate technology are the major challenges particularly in rural India. Creating various technological platforms and systems in States and UTs is also equally challenging. It also includes creation of digital infrastructure and digital interfacing.
- ***Organisational Challenges:*** These challenges include the systemic challenges, such as coordination and communication with regard to planning, implementation and monitoring of diverse programmes on digital education and digital learning.
- ***Economic Challenges:*** Digital India Programme also faces budgetary and financial challenges as implementation of the programme requires sufficient funding and financial support at central as well as State/UT level.

INNOVATIVE DIGITAL EDUCATION INITIATIVES & RURAL INDIA

Digital India and Digital Rural India require joining hands towards the nation building process by strengthening rural India through diverse interventions of 'Digital India' including 'Digital Education' and 'Virtual Learning'. Some of such initiatives are:

1. DIKSHA-Digital Infrastructure for Knowledge Sharing

Digital Infrastructure for Knowledge Sharing (DIKSHA) is a national initiative that provides access to a large number of curriculum linked e-content through several use cases and solutions such as QR coded Energized Textbooks (ETBs), courses for teachers, quizzes and others'. The salient features of DIKSHA include:

- Autonomy and choice within a national framework
- Online-offline and varied types of devices
- Diversity of content and energized textbooks
- Local language content and open licensing framework
- Diversity, flexibility and evolving

2. ePathshala

As a part of the digital march in education, ePhathshala has been a joint initiative of Ministry of Human Resource Development (MHRD), Government of India and National Council of Educational Research and Training (NCERT), New Delhi for the purpose of 'showcasing and disseminating all educational e-resources including textbooks, audio-video resources, periodicals and a variety of other digital resources'. Such an initiative facilitate facilitates people of rural India to access the print and non-print materials through websites and mobile apps leading to digital learning.

3. NROER-National Repository of Open Educational Resources

The NROER is a collaborative digital platform for sharing of open educational resources. As a joint venture of Ministry of Education, Government of India and CIET-NCERT, it hosts large number educational resources in many subjects and in different Indian languages for Primary, Secondary and Senior Secondary classes. Resources are available in different formats like Video, Image, Audio, Document and Interactive and to which the rural learners are open to access these resources.

4. e-Governance and Shaala Darpan

e-Governance is the efforts of reforming the country by reforming Government through technology. It is the use of technology to improve governance mechanisms and delivery of services. For example, Shaala Darpan is an e-Governance platform for all Kendriya Vidyalayas in the country including rural areas. It aims at improving quality of learning, efficiency of school administration, governance of schools & service delivery to key stakeholders.

5. e-Kranti: Electronic Delivery of Servies

As part of the digitalization process, Government of India has been making efforts for creation and maintenance of various digital platforms with technological interventions. The step towards providing education and other services through national e-Governance is a unique effort of Government of India for transparency and reliability of services. Even cyber security and other aspects of life and society have been covered under this initiative. Simply, e-Kranti is a march for rural transformation and reconstruction.

6. E-Granthalaya

E-Granthalaya is an Integrated Library Management Software developed by National Informatics Centre (NIC), Department of Electronics & Information Technology. The application is useful for automation of in-house activities of libraries and to provide various online member services. Rural India needs special focus towards this national initiative.

7. Digital Saksharta Abhiyaan (DISHA)

The Digital Saksharta Abhiyan or National Digital Literacy Mission (NDLM) Scheme has been formulated to impart IT training to people including Anganwadi workers, ASHA workers and authorised ration dealers in all the States/UTs across the country. It is an innovation for training, capacity building and resource pooling.

8. Pradhan Mantri Gramin Digital Saksharta Abhiyaan (PMGDISHA)

Pradhan Mantri Gramin Digital Saksharta Abhiyaan (PMGDISHA) in rural areas of our country is to make rural people digitally literate. It is to 'bridge the digital divide, specifically targeting the rural population including the marginalized sections of society like Scheduled Castes/Scheduled Tribes, Minorities, Below Poverty Line (BPL), women, differently-abled persons and minorities' *(Source: https://www.pmgdisha.in).*

9. **Shaala Siddhi**

A joint venture of NIEPA and Ministry of Education, Government of India is the National Programme on School Standards and Evaluation (NPSSE) which is known as Shaala Sidhdhi. It is a comprehensive instrument for school evaluation leading to school improvement. Shaala Siddhi enables the schools to evaluate their performance and thereby bring improvement. Rural India can use this mechanism for bringing transformations in school systems.

10 **Other Initiatives**

Some of other digital initiatives include Improving the Regulatory Framework, Bridging the Digital Divide, Driving Inclusive Participation, Mobile App Monitoring Systems, Child Tracking Systems, e-Sanjeevani, Electronic Vaccine Intelligence Network (eVIN), ICT in Education, Pedagogy of Digital Education, Swayam Prabha Channels, National Knowledge Network (NKN), Online Labs (OLABS), Project Management and Information System (PMIS), SMS-Based Mid-Day Meal Monitoring Scheme, Information for All, Universal Access and Mobile Connectivity, Public Internet Access System and many other.

Conclusion

India to be techno-savvy and knowledge power needs diverse technological interventions and innovative pedagogical practices. Digital education and virtual learning are the prerequisites of most of our rural development programmes. In this context, education is a nation-building process and digital education is the progressive education for building a healthy rural India, a well digitized rural India.

DR. RABINDRANATH LENKA
Director-Pedagogy, Research & Innovation
Centre of Research and Excellence for Advanced Technology in Education
CREATE-UNISED INDIA
Research Park, IIT Guwahati

Chapter 7

Inclusive Education - Inclusion of Children With Special Needs In Education

Education of children with disabilities, have led to progressive policy developments in India, and a surge in the enrolment of children who were traditionally excluded from the formal education system. This paper examines the perceptions and practices of mainstream teachers in rural government schools, within the context of increased learner diversity, focusing on how teachers understand, and respond to, the needs of children with disabilities. Data were collected through semi-structured interviews with teachers and classroom observations, in six primary schools, in three districts of Haryana. Our findings suggest that deficit-oriented views dominated teacher thinking, but they showed a readiness to engage with disability issues, recognising the value of education for all

However, they struggled in their classroom practices in relation to meeting diverse learner needs and exclusionary practices were further amplified for children with disabilities. Teachers were unwilling to take responsibility for the learning of children with disabilities, expressing significant concerns about their own preparedness, while highlighting the lack of effective and appropriate support structures. The paper concludes by drawing attention to the pressing need for effective teacher professional development opportunities and other support structures, to provide quality education.

INTRODUCTION

Free and compulsory education to all children up to the age of fourteen is the constitutional commitment in India. Despite spectacular quantitative expansion in every sphere of elementary education, the goal to achieve universal enrolment is still a far distant dream in a real sense. While adopting the constitution in 1950, the goal of UEE was to be achieved in ten years i.e. 1960. Keeping in view the educational facilities available in the country at that time, the goal of UEE was far too ambitious to achieve in a short span of ten years. Hence, the target date was revised several times.

During the decade 1991-2001, a number of Centrally Sponsored Schemes, as well as new programmes and projects were initiated across the Country. The Operation Blackboard (OBB) Scheme initiated in 1987 also got momentum during this period so as the large number of District Institutes of Educational Training (DIETs) established across the Country.

The Andhra Pradesh Primary Education Project (APPEP), Bihar Education Project (BEP), UP Basic Education Project, *Lok Jumbish* and *Shiksha Karmi* projects of Rajasthan and District Primary Education Programme (DPEP) were the main state-specific programmes initiated during 1991-2001. The mid-day meal scheme was also initiated during this period. Primary education remained the focus of all these programmes.
The DPEP which came to an end in 2000 was implemented in 272 districts across 18 states. The most recent ambitious programme, namely *Sarva Shiksha Abhiyan* (SSA) was launched by the Government of India in 2001 envisaged covering all non-DPEP districts before the end of the Ninth Five-Year Plan with a focus on the entire elementary level of education.

Initially, the focus of SSA was to further strengthen infrastructure in schools which was later shifted initially to increase enrolment (in terms of enrolment ratio, drop-out, and transition rate) and later focused more on improving quality of education. Under the aegis of SSA, a number of programmes were launched, and several studies were conducted (mostly by the *Technical Support Group* of SSA) in the beginning but later the momentum couldn't be maintained.

One of the significant visible achievements of SSA was strengthening the Educational Management Information System (EMIS) through DISE/U-DISE initiatives which had later acquired the status of the *Official Statistics* in 2012-13. Initially, the aim of SSA was to bring all the out-of-school, never enrolled, and dropped-out children back to school by 2003, achieve UPE by 2007, and UEE with satisfactorily Quality of Education by 2010 but nobody knows what had happened to all these targets, how much they achieved and what is the amount of the unfinished task.

Targets of 2003, 2005, and 2010 show the National commitment but no state-specific targets were set down in the process of which all states including Kerala and Bihar were given the same year to achieve the objectives of SSA. Neither district-specific target was set down. A thorough diagnosis exercise would have revel how many years a state and district would take to achieve the goal of UPE and UEE but the momentum of data analysis couldn't be maintained barring an initial period of SSA. Later the concept of *participatory planning* lost focus so as the development of district plans in a decentralized mode.

A number of capacity building programmes with an emphasis on the data analysis and use of indicators in planning were conducted across the county. Barring the initial period, later district plans started formulating at the State level and planning teams remained on paper. In most of the cases, Block Development Plans as envisaged in SSA couldn't be developed so as the School Development Plans based on which District plans were envisaged to be developed. In 2007, SSA extended to the secondary level of education in the form of *Rastriya Madhyamik Shiksha Abhiyan* (RMSA).

SAMAGRA SHIKSHA

Samagra Shiksha is a programme in the school education sector that extends from pre-school to class XII. This scheme is prepared with a broader goal of improving school effectiveness measured in terms of equitable learning outcomes and equal opportunities for schooling.

The Samagra Shiksha scheme contains the three Schemes of Sarva Shiksha Abhiyan (SSA), Teacher Education (TE) and Rashtriya Madhyamik Shiksha Abhiyan (RMSA). This scheme is launched for leveraging technology to manage and enhance efficient implementation of the centrally sponsored Integrated Scheme for School Education.

Under the Samagra Shiksha scheme, a 'PRABANDH' system was enabled on the Shagun Website, where the States and UTs can view the status of GoI (Government of India) releases, coverage as per UDISE (Unified District Information System for Education), approved outlays, school wise gaps, school wise list of approvals, cancellations in approvals, etc. Under the PRABANDH' system, the State or UTs can submit online monthly progress reports, financial and physical.

OBJECTIVES OF SAMAGRA SHIKSHA

- Providing quality education and enhancing the learning outcomes of students
- Bridging the gender and social gaps in school education
- Ensuring inclusion and equity at every level of school education
- Ensuring basic minimum standards in schooling provisions

- Promotion of vocationalisation of education
- supporting states to implement the right of children to compulsory and free education under the Right to Education Act, 2009
- To strengthen and upgrade State Institutes of Education or SCERTs and DIET as nodal agencies for teacher training

FEATURES OF SAMAGRA SHIKSHA: A HOLISTIC APPROACH TO EDUCATION

school education is treated holistically in perpetuity from Pre-school to Class 12. This scheme includes pre-school levels and senior secondary levels in school education for the first time.

ADMINISTRATIVE REFORM

A unified and single administrative structure leads to a harmonized implementation and flexibility to States for prioritizing their interventions is adopted under the scheme. This scheme aims at establishing an integrated administration system that looks at 'school' as a continuum.

FOCUS ON QUALITY OF EDUCATION

This scheme aims at enhancing and improving the quality of education by focusing on the two T's, i.e. Teachers and Technology and capacity building of the School Heads and Teachers. It supports Rashtriya Avishkar Abhiyan for promoting Maths and Science learning in schools and Padhe Bharat Badhe Bharat Programme for developing foundational skills at the primary level.

This scheme focuses on strengthening the Teacher Education Institutions like the District Institute for Education and Training (DIET) and State Council of Educational Research and Training

(SCERT) to improve the quality of prospective teachers. The nodal institution will be the SCERT for pre-service and in-service teacher training to make training need-based and dynamic. The Samagra Shiksha scheme will provide library grants for all schools ranging from Rs.5,000 to Rs.20,000.

FOCUS ON DIGITAL EDUCATION

The Samagra Shiksha scheme supports 'Operation Digital Board' in all secondary schools over a period of 5 years for ease in understanding and revolutionizing education. Technology-based learning classrooms to become flipped classrooms. This scheme enhances the usage of digital technology in education through digital boards, smart classrooms and DTH channels and strengthening digital initiatives like Shagun and UDISE+. It also aims to strengthen ICT infrastructure from upper primary to higher secondary level in schools

STRENGTHENING OF SCHOOLS

Under the Samagra Shiksha scheme, enhanced transport facilities to school children across classes from I to VIII for facilitating universal access to schools are provided. The composite school grant is increased from Rs.14,500-50,000 to Rs.25,000-Rs.1,00,000 allocated on the basis of school enrolment. This scheme makes specific provision for 'Swachhta' activities to support 'Swachh Vidyalaya' and improve the infrastructure quality in Government Schools

FOCUS ON GIRL EDUCATION

The Samagra Shiksha scheme upgraded Kasturba Gandhi Balika Vidyalayas (KGBVs) from Class 6-8 to Class 6-12. It provides self-defence training for girls from the upper primary to the senior secondary stage and commitment to the 'Beti Bachao Beti Padhao' initiative

FOCUS ON INCLUSION

Under the Samagra Shiksha scheme, uniforms allotted under the RTE Act is enhanced from Rs.400 to Rs.600 per annum per child and textbooks enhanced from Rs.150/250 to Rs.250/400 per annum per child. The energised textbooks are introduced. The allocation for Children with Special Needs (CWSN) is also increased from Rs.3,000 to Rs.3,500 per annum per child. A stipend of Rs.200 per month for girls with special needs is provided from Classes 1 to 12

FOCUS ON SKILL DEVELOPMENT

The exposure to Vocational Skills at the upper primary level would be extended under this scheme. The vocational education for Class 9 to Class 12 as integrated with the curriculum to be made more industry-oriented and practical. This scheme will reinforce the emphasis on the 'Kaushal Vikas' scheme.

FOCUS ON SPORTS AND PHYSICAL EDUCATION

Sports Education will be an integral part of the curriculum. All schools will receive sports equipment at the cost of Rs.5,000 for primary schools, Rs.10,000 for upper primary schools and Rs.25,000 for secondary and senior secondary schools for inculcating and emphasizing the relevance of sports.

FOCUS ON REGIONAL BALANCE

This scheme will promote balanced educational development and provide preference to the LWE affected districts, Educationally Backward Blocks (EBBs), Border areas, Special Focus Districts (SFDs) and the 117 aspirational districts which are identified by Niti Aayog.

BENEFITS OF SAMAGRA SHIKSHA

The Samagra Shiksha scheme will help improve the transition rates across different school education levels and aid in promoting universal access to children for completing school education. The integration of Teacher Education will facilitate effective linkages and convergence between various support structures in school education through interventions such as innovations in pedagogy, unified training calendar, mentoring and monitoring, etc. This scheme will enable school education to reap the benefits of technology and widen access to quality education across all States and UTs, and all sections of society.

INCLUSION OF CHILDREN WITH SPECIAL NEEDS IN SAMGRA EDUCATION

The NPE, 1986 and POA, 1992 gives the basic policy framework for education, emphasizing on correcting the existing inequalities, it stresses on reducing dropout rates, improving learning achievements and expanding access to students who have not had an easy opportunity to be a part of the general system. The NPE, 1986 envisaged some measures for integrating of children with physical and mental handicap with the general community as equal partners, preparing them for their normal growth and development and enabling them to face life with courage and confidence.

India has also been signatory to international declarations like the Salamanca Statement and framework for action on special needs education (1994) and Biwako Millennium Framework for Action (2002) and the UN Convention on the Rights of Persons with Disabilities, 2006 that emphasize the need for fundamental educational policy shifts to enable general schools to include children with disabilities. The NCF-2005 recommends making the curriculum flexible and appropriate to accommodate the diversity of school children including those with disabilities in both cognitive and non-cognitive areas.

The Centrally Sponsored Scheme of Integrated Education for the Disabled Children was introduced in 1992 with a view to providing educational opportunities for children with disabilities in general schools, to facilitate their retention in the school's system. It provided for facilities to students with disabilities including expenses on books and stationery, expenses on uniforms, transport allowance, reader allowance, escort allowance, hostel accommodation and actual cost for equipment's. The scheme also supported the appointment of special teachers, provision for resource rooms and removal of architectural barriers in schools.

The Centrally Sponsored Scheme of SSA of 2001 had set time-bound targets for the achievements of UEE. With "zero rejection' as its cornerstone, the programme provided support for the inclusion of children with disabilities in general schools at the elementary level. The SSA norms were further strengthened by the RTE Act, 2009 which talks about free and compulsory elementary education to children without any discrimination on the basis of caste, gender, disability etc.

While the RTE Act mandates inclusion of CWSN, some may be unable to attend school despite specific interventions designed for their education. The amendment of RTE Act (in August 2012) has included CWSN in the definition of disadvantaged groups. It includes children with severe - multiple disabilities with the right to opt for Home Based Education (HBE), thus creating an enabling environment for all children.

As SSA supported inclusion of children with special needs at the elementary education level, a need was felt for a scheme for the disabled children at secondary stage. The IEDSS scheme was, therefore implemented to enable all children and young persons with disabilities to have access to secondary education and to improve their enrolment, retention and achievement in the general education system. Under the scheme every school was proposed to be made disabled-friendly.

The Centrally Sponsored IEDSS Scheme aimed at enabling all students with disabilities completing eight years of elementary schooling an opportunity to complete four years of secondary schooling (classes IX to XII) in an inclusive and enabling environment and provided educational opportunities and facilities to students with disabilities in the general education system at the secondary level (classes IX to XII).

Samagra Shiksha aims to look at education of all children including CWSN in a continuum from pre-school to class XII. The scheme will cover all children with special needs with one or more disabilities as mentioned in the schedule of disabilities of the Right of the Persons with Disabilities (RPWD) Act, 2016 studying in Government, Government-aided and local body schools.

The scheme stresses on working in convergence with all the line Departments/ Ministries and intends to provide relevant holistic support for effective and appropriate services for education of CWSN.

The objectives of the scheme are:

- To enable all children and young persons with disabilities to have access to inclusive education and improve their enrolment, retention and achievement in the general education system.

- Identification of children with disabilities at the school level and assessment of her/his
- educational needs.

- Provision of aids and appliance and assistive devices, to the children with special needs as per requirement.

- Removal of architectural barriers in schools so that students with disability have access to classrooms, laboratories, libraries, and toilets in the school.

- Supplying appropriate teaching learning materials, medical facilities, vocational training support, guidance and counselling services and therapeutic services to children with special needs as per their requirement in convergence with line departments.

- General school teachers will be sensitized and trained to teach and involve children with special needs in the general classroom. For existing special educators, capacity building
- programmes will be undertaken.

- CWSN will have access to support services through special educators, establishment of resource rooms, vocational education, therapeutic services and counselling.

Target Group

The scheme will cover all children from pre-school to senior secondary stage studying in Government, local body and Government-aided schools, with one or more disabilities as defined under the Rights of Persons with Disabilities Act (2016) namely:

1. Blindness
2. Low-vision
3. Leprosy Cured persons
4. Hearing Impairment (deaf and hard of hearing)
5. Locomotor Disability
6. Dwarfism
7. Intellectual Disability
8. Mental Illness
9. Autism Spectrum Disorder
10. Cerebral Palsy
11. Muscular Dystrophy
12. Chronic Neurological conditions
13. Specific Learning Disabilities
14. Multiple Sclerosis
15. Speech and Language disability\
16. Thalassemia
17. Hemophilia
18. Sickle Cell disease
19. Multiple Disabilities including deaf blindness
20. Acid Attack victim
21. Parkinson's disease

Girls with disabilities will receive special focus and efforts would be made under the scheme to help them gain access to schools, as also to provide motivation and guidance for developing their potential. All the enrolled CWSN will be covered through UDISE/SDMIS and will eventually be linked with Aadhaar, ultimately facilitating the State/UT to track CWSN of both systems i.e. those that are in school and those with Home based. The UDISE will have the relevant details of children. Further an extensive database will be developed which will cover all the particulars of children including the type of disability, degree of severity, medical needs, emergency contacts and all other relevant details that will help the school management to cater to the needs of the CWSN.

Components of the Scheme

The Scheme will include assistance for two kinds of components:

Student oriented component

The student oriented component may be utilized for specified items like: -

(i) Identification and assessment of children with disabilities. Provision of aids and appliances, medical services, diagnostic services etc.

(ii) Access to and development of teaching learning material as per requirement etc.

(iii) Provision of facilities like transport/escort facilities for children with severe disabilities, hostel facilities, scholarship, assistive devices, support staff (readers, amanuensis) etc.

(iv) Provision of large print textbook, Braille books and uniform allowance.

(v) Stipend for Girl Students @ Rs. 200 per month for 10 months to encourage their participation in the school system for all girls with disabilities studying in Classes-I to XII.

(vi) The use of ICT to increase access to a vast amount of information not otherwise available.
(vii) Awareness and motivation camps for increasing enrolment and retention.

(viii) Provision of Home-Based Education for children with severe multiple disabilities with a view to mainstream them in the general education system.

Resource support

For providing resource support to CWSN, the existing human resource appointed under SSA and RMSA will be rationalized, and the remaining vacancies may be filled through fresh appointment as per the norms provided at All special educators should be registered with Rehabilitation Council of India (RCI). These educators should mandatorily be available for all CWSN including the children with high support needs as well.

The educators may be posted at the block or cluster level or as per the requirement and can operate in an itinerant mode, covering a group of schools where children with special needs are enrolled so that each child with special need is adequately covered.

Education of teachers and other stakeholders

Intensive teacher education programme will be undertaken to sensitize and build capacity of the regular teachers and resource teachers on meeting learning needs of all teachers to provide quality education to CWSN and improve their learning outcomes. This teacher education programme will be recurrent at block/cluster levels and integrated with the on-going in-service teacher education/training schedules in DIETs and other Institutions.

Teacher education modules at SCERT, DIET and BRC level should include suitable components on education of children with special needs. Training of educational administrators including headmasters, all other staff & relevant personnel of school education should be regularly organized.

Research and Development

The Integrated Scheme will encourage research and development activities in all areas of education of children with special needs including action research, research to improve learning of CWSN, especially focusing on children with high support needs (Thalassemia, Hemophilia, Sickle Cell disease and Chronic neurological conditions etc), eventually leading to concrete learning outcomes. For this, convergence with different Ministries like Ministry of Health and Family Welfare, MoWCD, Ministry of Social Justice and Empowerment, Ministry of Sports and Youth Affairs, private organizations with Corporate Social Responsibility (CSR) funds etc.

Social Access to CWSN

Ensuring social access to CWSN is a greater challenge as compared to providing physical access as it requires an in-depth understanding of the various educational needs of CWSN. A very important dimension of social access is discrimination. CWSN are subjected to many forms of discrimination. Teachers and peers have a very important role to play in this context. Social access could include the following:

i. Parental training
ii. Peer sensitization
iii. Awareness building of different stakeholders across all levels
iv. Special emphasis will be given to education of girls with disability

Monitoring & Evaluation:

On-going monitoring and evaluation should be carried out to refine the programme from time to time. For this, appropriate monitoring mechanisms and tools are to be devised at every level and field tested at regular intervals.

Conclusion:

Although, the scheme tried to integrate various components of Rights of Persons with Disabilities Act 2016, the provision of the scheme provides glossy picture as provisions are in form of tokenism rather than a wholehearted attempt to provide quality education for Children with Disabilities/Children with Special Needs as per the RTE, 2009 and RPWD Act 2016.

First and foremost there has been no parity maintained in the entire scheme in the usage of terminologies. Nomenclature like Children with Special Need, Children with Disability, handicapped and Divyaang has been used interchangeably.

The scheme comes after the enactment of a new law on the rights of persons with disability, yet it to fails to use and define various terminologies and nomenclature of RPD – reasonable accommodation, communication, augmentative and alternative mode of communication, barrier-free both in words and in principles. The scheme fails to define high support children and other terminologies. Additionally, the scheme also fails to effectively interrogate with the new terms and concepts in the law. Defining these categories is essential for proper allocations of the services and entitlement to these children.

Exclusion from definition, ambiguities related to the terminologies and adopting broad definition leaves sufficient scope for stakeholders to bypass the provisions.

The scheme has largely remained silent and does not engage on the mechanism of knowledge building on the 21 new impairment groups.

Additionally, the term inclusive education, the main thrust area of the scheme has been used with different meaning in different chapters. While the scheme in its objective and goals incorporate the inclusive education for children with disabilities along with other categories of the disadvantageous section this parity is not maintained. The organization of the various chapters in the scheme reflects this divergence as inclusive education of girls and children of other traditionally disadvantageous and marginalized section are dealt in one common chapter; a separate chapter has been designed for the education of Children with Disabilities/Children with Special Needs. Such segregation within the text of the scheme leads to failure of the core objective of the scheme.

The scheme has little scope for the children with high support and home-based learning as it fails to provide any provisions and modalities to make education accessible to them. The financial norms of the scheme specify that the key thrust of the scheme will be on providing inclusive education to all Children with Disabilities/Children with Special Needs in GENERAL SCHOOL thus sidelining those categories of students that are under the home-based learning or special schools.

There are provisions for formulation and implementation of new scholarship scheme for the financial assistance for these children except stipend of Rs. 200 for 10 months for girls with special needs. Ambiguities prevail over extending this stipend at all level of school education.

Last but not the least, the policy-makers have failed to address the issue of child protection and safety inside the school premises in a comprehensive manner. The entire scheme does not provide any detailed provisions, mechanism or guidelines for the safety and protection of the children in general and Children with Disabilities/Children with Special Needs in particular within the premises of the schools given their vulnerability to all forms of physical and sexual violations. Also, one could not find any mention of Disability Commissioner, its role, function and its involvement with National Commission for Protection of Child Rights or with the State Commission for Protection of Child Rights in case of violation of their educational rights.

DR. DAYAL PYARI
Assistant Professor, AIE
(Amity University, Noida, UP)

Chapter 8

Image Segmentation

Image segmentation is the process of partitioning an image into multiple segments. It is typically used to locate objects and boundaries in images. The basic architectures contain the encoding and decoding networks with the activation function for the normalization.

There are many segmentation architectures which uses the machine learning and computer vision algorithms like Fast FCN network, U-Net, Gated SCNN etc, these segmentation architectures are used mostly in the biomedical images for detection of the abnormality and for the detection of the disease.

The U-Net architecture is one of the image segmentation architectures which is first introduced in 2015 it has shape of the U which is used for the image segmentation and its training architecture contains the encoder and decoder subnets with the paths which are contracting paths and expanding path including the ReLU activation function.

This paper discuss about the U-Net architecture and the other novel architectures which are based on the U-Net architectures and use the applications of the U-Net architectures with their advantages and disadvantages, and hence we can tell why U-Net architecture is important in area of biomedical imaging and for the object detection in the aerial and landscape images.

Introduction

A digital image is made up of various components that needs to be "analysed", let's use that word for simplicity sake and the "analysis" performed on such components can reveal a lot of hidden information from them. This information can help us address a plethora of business problems – which is one of the many end goals that are linked with image processing.

Image-Segmentation is the process by which a digital image is partitioned into various subgroups (of pixels) called Image Objects, which can reduce the complexity of the image, and thus analyzing the image becomes simpler

[1]. There are various image segmentation architectures based on machine learning algorithms. Some of these are: Fast Fully Connected Network (Fast FCN)

[2]. architecture, Convolutional Neural Networks (CNN) based encoder-decoder networks

[3]. Gated Shape CNN (Gated SCNN) architecture and U-Net Architecture

[4]. U-Net Architecture is evolved from the traditional Convolutional Neural Networks (CNN)

[5]. It was first designed and implemented in 2015 for the processing of the biomedical images.

[6]. A CNN is focused on image classification, where input is an image and output is one label, but in the biomedical image segmentation, it is required not only to distinguish whether there is a disease but also to localize the area of the abnormality.

[7]. The network consists of two paths contraction path and expansion path. The contraction path consists of the repeated

application of two 3x3 convolutions. The training of encoding sequence is done by VGG-11 encoder. In the segmentation tile 3x3 convolution layers are added with rectified linear unit (ReLU) activation function. A max pooling operation of 2x2 convolutions is done at each layer for down-sampling process.

[8]. The number of feature channels is doubled in each down-sampling step. Every step in the expansion path consists of an up-sampling of the feature map followed by a 2x2 convolution ("up-convolution") that halves the number of feature channels. The cropping is necessary due to the loss of border pixels in every convolution. At the final layer a 1x1 convolution is used to map each 64-component feature vector to the desired number of classes. In total the network has 23 convolutional layers.

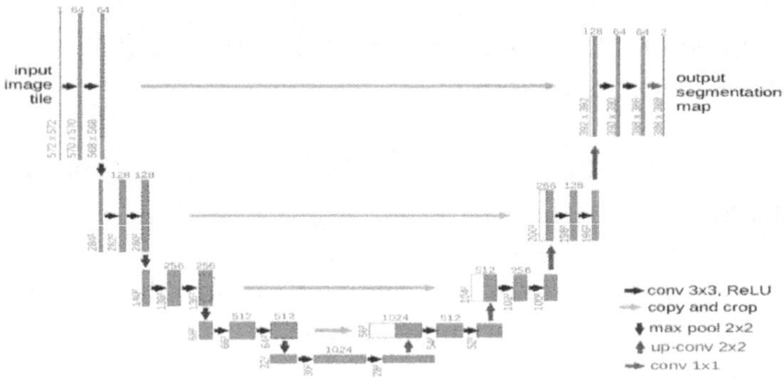

Figure-1.1 shows the basic U-Net Architecture .

The basic U-Net architecture is the network of plain skip connections

[9]. The layers are connected directly to each others in this network. The skip connections are extra connections between nodes in different layers of neural networks that skip one more layers of non-linear processing. Dense skip connections.

[10]. Fully Connected Connection, are a type of layer in a deep neural network that use a linear operation where every input is connected to every output by a weight. Full scale skip connections are the hybrid approach of the skip connections. It is the combination of dense connections and plain skip connections.

Rest of the paper is organized as follows: Section-2 contains the survey of several research works done on U-Net based architectures. The comparative study of different proposed architectures on the basis of different parameters is discussed in section 3. Section-4 presents our observations from the study of the different architectures based on U-Net. Section-5 concludes the paper.

1.Literature Survey

Several researchers have proposed various models based on the basic U-Net architecture [11-20]. The U-Net based architecture and its variants can be classified on the basis of plain skip connections, dense skip connections and full-scale skip connections [8, 9, 10].

In 2015, Olaf Ronneberger *et al.* developed the first U-Net architecture using CNN with plain skip connections for image segmentation [5]. It connects the encoding and decoding layers without any interference. The training of encoding sequence is done by VGG-11 encoder. The 3X3 convolution layers are added in the segmentation tiles. For the decoding 2X2 up-sampling layers are added in the architecture.

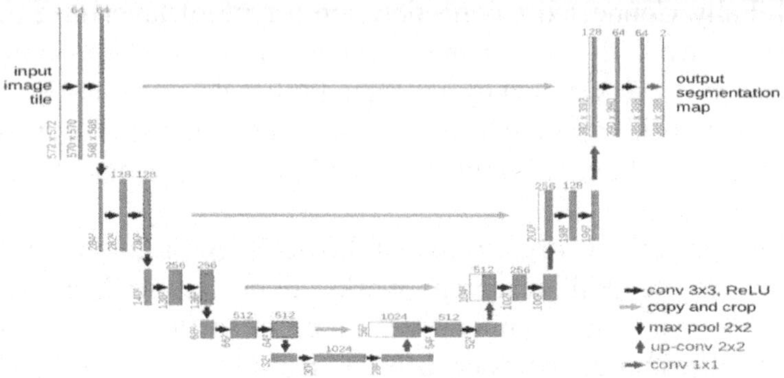

Figure 2.1: The Basic U-Net architecture

It takes the input as full sized images and shows the labeled output and hence it is easy to find the abnormality. Basic U-Net architecture can be easily scaled to have multiple classes. It is relatively easy to understand and suitable to work with the convolution process. It works well with small training sets

The size of U-Net must be comparable with the features included otherwise result could be ambiguous. The U-Net architecture takes significant amount of time to train. It requires relatively high GPU memory unit for larger images.

In 2017, Fabian Isensee *et al.* proposed a robust segmentation algorithm in the form of CNN inspired from U-Net architecture. This network is designed to process the large 3D input blocks of 128x128X128 voxels. The network comprises the context aggregation pathways. It encodes increasingly abstract representations of inputs as we go deeper in the network. The activations in the context pathway are computed by the context modules. These context modules are pre activation residual blocks. The figure 2.1 shows the architecture which can be used for the segmentation of the larger 128X128X128 3D voxels.

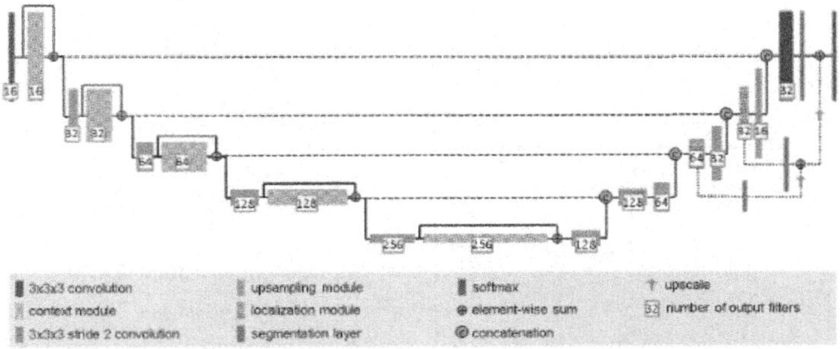

Fig-2.1: U-Net architecture for the 3D 128x128x128 voxels segmentation

It was the robust segmentation network based on U-Net which can process large 3D inputs. It follows the concatenation and recombines the features, which reduces the critical memory consumption. The five-fold cross validation was applied to the training cases which minimizes the mean square error. The network requires high GPU memory unit to process large 3D input. It also takes significant amount of time to train.

In 2018 Debesh Jha *et al.* has proposed the novel architecture named Double U-Net. It is a combination of two U-Net stacked on the top of each other. Max Pooling operation is performed with 2X2 window which reduces the spatial dimension of the feature maps. The decoder performs 2X2 bi-linear up-sampling on input features. It doubles the dimensions of input feature maps. The output is generated by concatenation process followed by two 3X3 convolution operations.

Figure 2.3 shows the architecture of Double U-Net [12].

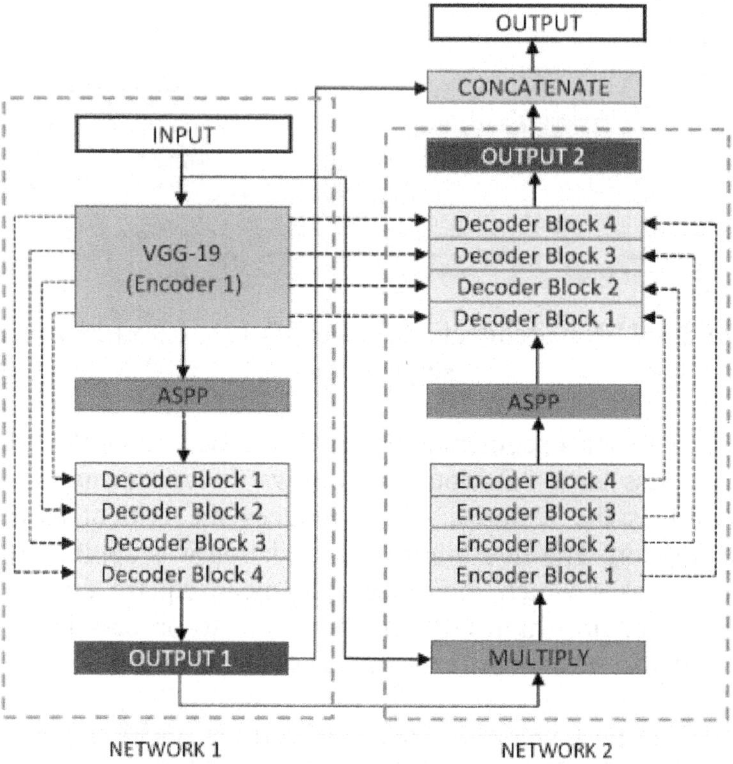

Figure 2.3: Double U-Net Architecture [12]

This architecture is an approach to improve the performance of U-Net. It uses the pre-trained VGG-19 which has learned features. It easily outperforms U-Net in segmentation and capturing semantic information. It is possible to integrate more CNN blocks in Double U-Net. The network may require a powerful GPU to process the input. It uses more parameters than U-Net which leads to an increase in training time.

In 2019 Reza *et al.* proposed a model named Double U-Net Architecture Bi-directional convLSTM U-Net. This architecture used the sandwich of two U-Net networks with the convLSTM neural network which is used to train the model, it is a recurrent neural network which has spatio temporal prediction property. This architecture is specifically used in convolutional neural network predictions. The figure below shows the architecture of Double U-Net architecture with the ConvLSTM also Bi-directional Convolutional LSTM U-Net architecture.

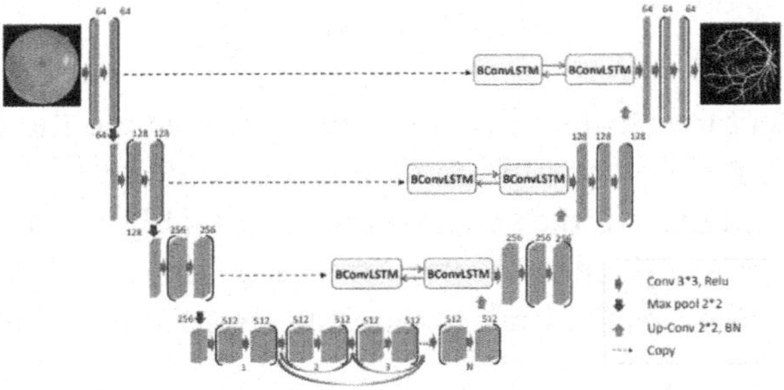

Figure 2.4: BCDU-Net with bi-directional ConvLSTM in the skip connections and densely connected convolution.

This architecture has used the functionality vgg16 encoder and added three vgg16 encoders along with the recurrent neural network and the decoders are added for the down sampling process and thus, generates the output in the form of image.

In 2020 Fabian Isensee *et al.* proposed nnU-Net for the brain tumor segmentation and model took the first place in the BraTS 2020 competition the nnU-Net architecture is combination of three basic architectures U-Net, cascaded U-Net and 3d U-Net. The following image shows the figure of nnU-Net architecture.

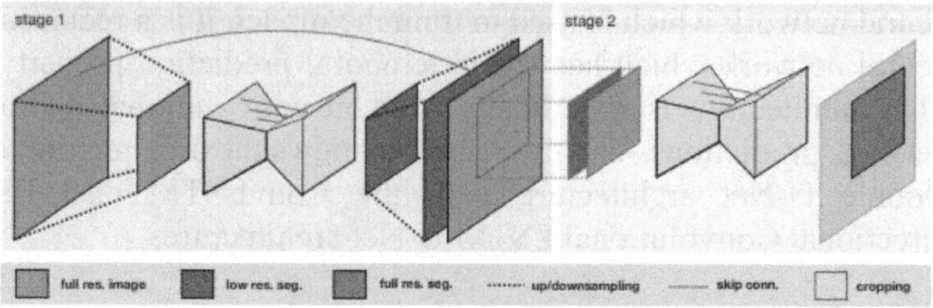

Fig 2.5: nnU-Net architecture, the combination cascade U-Net and 3D U-Net.

This architecture works in two stages at first stage the cascade U-Net takes the high resolution image and converts that into low resolution and then in the second stage the 3-D U-Net takes that low resolution image and performs the up/down sampling processes and gives the high resolution segmentation output.

The table below summarizes the previously proposed models and their findings.

Table-1 Literature Survey

Authors and Year	**Proposed Model**	**Findings**
Olaf Ronneberger *et al.* 2015 [3]	Proposed the first basic U-Net architecture with the help of Convolutional Networks for the segmentation of the biomedical images.	The basic U-Net architecture for the segmentation of the images with dice score of 0.7756, PhC score of 0.9203.
Fabian Isensee *et al.* 2017[2]	Proposed model is based on the U-Net architecture which can do the survival prediction by the help of the image segmentation using U-Net architecture.	Authors have trained and evaluated their architecture using BraTs 2017 and 2015 dataset.It has the dice score of 0.80 and the accuracy score of survival prediction is 0.86.

Debesh Jha *et al.* 2018 [6]	Proposed a novel architecture inspired by basic U-Net architecture called Double-U-Net architecture, whichis a combination of two U-Net architectures stacked on top of eachother.	The experiments were conducted on MICCAI 2015 and CVC clinicDB. The experiments on double U-Net has dice score of 0.9239, while basic architecture has score of 0.8781 with the precision of 0.9592.
Zongwei Zhou *et al.* 2018 [11]	Proposed a model called U-Net ++ for the biomedical image segmentation, which is a deeply-supervised ncoder-decoder based network where the encoder and decoder sub-networks are connected through a series of nested, dense skip pathways. The re-designed skip pathways aim at reducing the semantic gap between the feature maps of the encoder and decoder sub-networks.	the model is trained and evaluated on the MICCAI 2018 and LIDC dataset and gave the precision of 77.21 while the basic architecture gave the precision of 71.67, which is better result that basic architecture.
Leonid Ivanovsky*et al.* 2018 [13]	Proposed a methodology for segmentation of aerial images on the cityscape dataset for the building detection with the help of the U-Net Architecture.	Used the model for the application of building detection with application of LinkNet with the basic U-Net architecture, the model is trained on the Spacenet and city landscape dataset. It has the accuracy of 96.31% with the dice score of 0.77.

Simon A. A. Kohl*et al.* 2018 [14]	Proposed a generative segmentation model based on a combination of a U-Net with a conditional variational autoencoder that is capable of efficiently producing an unlimited numberof plausible hypotheses. They show on a lung abnormalities segmentation task and on a Cityscapes segmentation task that their model reproduces the possible segmentation variants as well as the frequencies with which they occur, doing so significantly better than published approaches.	. The model is trained and evaluated on the LIDC-IDRI dataset, the model gives 1.5% accurate result while the basic architecture only gives 0.5 % accuracy on LIDC-IDRI dataset.
Reza Azad*et al.* 2019 [7]	Proposed an extension of U-Net, Bi-directional ConvLSTM U-Net with Densely connected convolutions(BCDU-Net), for medical image segmentation, in which they use full advantages of U-Net, bi-directional ConvLSTM(BConvLSTM) and the mechanism of dense convolutions	,The given model has tested and evaluated on the ISIC and Lung segmentation dataset 2015, the BCDU-Net
Fabian Isensee *et al.* 2020 [4]	Proposed a model nnU-Net for the brain tumor segmentation and model took the first place in the BraTS 2020 competition The nnU-Net architecture is combination of three basic architectures U-Net, cascaded U-Net and 3d U-Net. It works in two stages in stage-1 the cascade U-Net takes full res. Image	The experiments are done on the BraTs 2020 dataset with the dice scores of 88.95, 85.06 and 82.03, and HD95 valuesof 8.498,17.337 and 17.805 for whole tumor, tumor core and enhancingtumor, respectively.

	and convert it into low res segmentation. The added 3d U-Net takes that low res. Segmentation and performs the up/down-sampling operations and gives the high res. segmentation output label.	
Jeya Maria Jose Valanarasu*et al.* 2020 [5]	Proposed model called KiU-Net based on the basic architectureis a 3D convolutional architecture for volumetric segmentation. This model named KiU-Net uses the application of Kite-Net with U-Net architecture	, It has dice score of 0.7321 while basic architecture's dice score is 0.6711 on GLaS dataset.
Peng Zhao *et al.* 2020 [12]	Proposed a deep learning image semantic segmentation network named Spatial-Channel Attention U-Net (SCAU-Net) for the segmentation of the medical images. SCAU-Net has an encoder-decoder-style symmetrical structure integrated with spatial and channel attention as plug-and-playmodules. The main idea is to enhance relative features and restrain irrelevant features at the spatial and channel levels.	The model is trained on the GLaS and CRAG dataset and has a dice score 0.9063 and 0.9100 while basic architecture has score of 0.8963 and 0.9003.
Xuebin Qin*et al.* (2020)[15]	Proposed a model named U^2 Net architecture which is based on the nested U-Net architecture and the presented model is combination of two nested U-Net architectures.	The experiments are done on the ECSSD and SOD dataset the results on the dataset are improved than the basic unit. The dice score of the basic U-Net is 0.896 and the dice score of the given architecture is 0.951.

Huimin Huang*et al.*(2020)[16]	Proposed a model named UNet 3+ architecture which is combination of basic UNet and UNet++ architecture. It is a full scale skip connection architecture which has the properties of U-Net architecture(plain skip connection) and U-Net++(the nested connection). Thus it is a scaled architecture of U-Net and U-Net++.	The experiments are done on the ISBI LiTS 2017 datsets the comparisons done on the basic UNet architecture, UNet++ and UNet3+. The dice score of the basic UNeton the given dataset is 0.9332, the dice score of UNet++ 0.9352 and Unet3+ has score of 0.9552.

2.Comparative Studies of the architectures

The architectures given by the researchers above are efficient and effective, the given models has many similarities and there are many techniques included to improve the performance of the architecture and the models proposed by them, for example:

- All the given models use the encoders and decoders.
- All the given models follow the up-sampling and down-sampling process.
- All the architectures use the activation functions like ReLu and leaky ReLu.
- All the architectures use the applications of convolutions.

The following table shows the comparison of the architectures and what makes each one of them different from each other;

Author and Year	Architectures	Encoders/decoders number	Layers added	Added architectures	Up/down-sampling numbers	The major difference	Strengths	Weakness
Olaf Ronneberger *et al*. 2015 [3]	Basic U-Net Architecture	At each step there are 3 VCG-11 encoders and 3 decoders.	5 encoding and decoding layers.	There are no added architectures and this basic one uses the application of CNN.	5 layers of 3X3 conv used in upsampling and down-sampling.	1. The most basic architecture, which is later added in the other improved models as reference.	One of the most efficient architecture used with the CNN and convolution.	It is a very slow model and requires much gpu for the process which requires much time to train the model.
Fabian Isenseeet *al.* 2017[2]	Survival prediction using UNet	The 3 VCG-11 encoders and decoders used.	128X128X128 voxels added with 5 layers in architecture.	Used basic unet architecture for survival prediction in the model.	5 layers with 3X3X3 conv. Up-sampling and 1X1X1 conv. Down-sampling.	1. The architecture uses basic unet with the improved conv 3x3x3 and 1x1x1 conv feature maps.	It uses the basic architecture in an effective way of convolutions, with improved predictive modeling	The only weakness of this survival prediction model is that it can give overfitted results.
Fabian Isenseeet *al.* 2020 [4]	nnU Net architectures	Used 3 encoders for 3d unet and 3 decoders for the 3d images inputs, for 3d unet .	5 layers like original model with added residual connection and attention mechanism.	The added architecture is original UNet architecture added with the attention mechanism and residual conection.	3d unet layers for down-sampled data which results segmentation maps which up-sampled from the original resolution.	1. It is basic Unet with the residual connection and attention mechanism Which can do the segmentation of large inputs like 482x482x512 to 256x256x256 output.	It can perform the segmentation on the large size of input and can perform better with 3d unet and can give full upsampled resolution.	It can give overfitted results and sometimes can do the segmentation of the unnecessar-y areas of images.

Author and Year	Architectures	Encoders/decoders number	Layers added	Added architectures	Up/down-sampling numbers	The major difference	Strengths	Weakness
Jeya Maria Jose Valanarasu *et al.* 2020 [5]	KiU Net Architecture	Total 3 number of encoder decoder used in U-Net and 3 encoder decoders added in the Ki-Net architecture and thus this is the architecture the combination of two architectures.	Used total of 6 layers where 3 layer are for conv2d+up-sampling+ReLU and 3 layers are for conv2d+maxpooling + ReLU and similarly for the 3d network.	Combination of two architectures the U-Net architecture added with the new Ki-Net and other architectures seg-Net, Res-Net and U-Net++ are also added and for 3d architecture the 3D-Unet is used with 3D-Ki-Net.	3 layers are added for the up-sampling+conv2D+ReLU and 3 layers are added for the maxpooling +conv2D+ReLU /down-sampling.	1. Unlike other architectures, it's the only architecture which is used for 2D and 3D convolutions. 2. this architecture can work on the images which are of larger size with dice score of 0.8431 and jaccard score of 0.7422 for the dense KiU-Net.	Can work on dense networks and it can do the segmentation of the larger images and it is the most scalable model which can be used as KiU-Net, 3D KiU-net and dense KiU-Net.	Since it uses many architectures as parameters and because of its scalability and dense connection, it requires larger GPU and high performing platform.
Debesh Jha *et al.* 2018 [6]	Double U-Net Architecture	The encoder-1 is VCG-19 is used in first U-Net and 4 decoder blocks. In the second U-Net there are 4 encoder and decoder blocks used in the given architecture.	Total 3 layers of encoder /decoder used in first U-Net and in the second U-Net there are 5 layers including output and concatenation layers in the architecture.	Combination of two consecutive U-Net architectures including the encoder and ASPP model which uses the applications of the fully connected convolution networks.	3 layers are added for the up-sampling in the first U-Net, it performs two 3x3 convolutions followed by batch normalization, after the up-sampling there is 2x2 maxpooling, similarly goes with the second U-Net.	1. This architecture is a scaling of the basic U-Net architecture it performs precisely than the other nested and basic architecture. 2. strength of this architecture, it uses encoders/ decoders, it uses powerful VCG-19 encoder in the first U-Net and there are four encoders added with ASPP network.	It is a powerful model and can give much precise results because of its powerful encoding/ decoding sequence and gives precision of 0.9133 dice score.	Since it uses the combination of two basic U-Nets the major weakness of this model is that it requires too much time for the training and high performing GPU.

Author and Year	Architectures	Encoders/decoders number	Layers added	Added architectures	Up/down-sampling numbers	The major difference	Strengths	Weakness
Reza Azad *et al.* 2019 [7]	Bi-directional convLSTM U-Net	The encoding path use the 3 VCG-16 encoder with recurrent neural network and the decoders are made from the scratch in the model. The encoding path contains 3x3 filters followed by 2x2 max pooling functions added with ReLU and in decoding path 2x2 convolution applied.	There are total of 4 layers and in each layer there is densely connected convolutions, in each encoding path there are 3x3 conv with ReLU and in decoding path 2x2 max pool with 2x2 up-conv with Bi-directional convLSTM.	The architecture uses basic U-Net with the densely connected convolution networks and added the recurrent neural network and CNN.	For the down-sampling there is 3x3 conv with ReLU activation function, and after the encoding the process of up-sampling in decoding path follows 2x2 convolutions, it doubles each feature maps followed by batch normalization	1. The major strength of this model is the application of LSTM with bi-directional convolution networks and it uses the application of RNN with densely connected neural networks. It makes the model works more precisely than others.	The strength of the model is densely connected convolutional networks with the applications of basic U-Net model and it makes it work better than the basic architecture with preciseness of 0.851.	The most common weakness of this model is training duration, the dense connection of neural networks takes time to train the model after every batch normalization and it makes it slower than the others.

Author and Year	Architectures	Encoders/decoders number	Layers added	Added architectures	Up/down-sampling numbers	The major difference	Strengths	Weakness
Zongwei Zhou *et al.* 2018 [11]	U-Net++ a nested U-Net architecture	The model consists of the series of the nested dense U-Net architectures where each block consist of encoder /decoder pairs. Each convolution in the model used to bridge the gap between encoder /decoder.	There are nested layers of unet blocks and each block consists of three convolution layer. In the model the initial layer is called backbone which is used for basic up/down-sampling, while the other layers are nested blocks used to bridge the semantic gaps.	The Unet++ architecture is nested model consists of nested basic U-Net models represents the blocks densely connected with the others, thus this model becomes too complex than the others, in the model there are two basic layers for up and down sampling, the other layers are complex and they are used to for inference and semantic bridge.	Unlike other models given, the U-Net ++ only has a backbone layer for the up/down-sampling, and the other layers are used to train that layer, the layers are using the 3x3x3 convolution for 3d images segmentation and nuclei training.	1. Unlike other models U-Net++ is the model which use the application of the nested U-Net architectures where many unets are used in the form of blocks and all the blocks are connected to each other in a nested way and down-sampling happens in the backbone layer. Hence it is much more scalable than the other models and can be used in another model too.	It is a nested U-Net model which uses the application of U-Net models connected to each other and the connected models are always open to add some other models which makes it more scalable and gives precise results than other models.	The nested U-Net model is very efficient and can be used for the segmentation of the larger images since it is nested it requires too much time to train the models.

Author and Year	Architectures	Encoders/decoders number	Layers added	Added architectures	Up/down-sampling numbers	The major difference	Strengths	Weakness
Peng Zhao *et al.* 2020 [12]	Spatial-Channel Attention U-Net (SCAU-Net) Architecture	There are total 5 numbers encoder /decoders which contains spatial attention, block(64) and input in the initial layer, and in the decoder layer spatial channel, block(64) and unspooling blocks are added before conv(1x1,c) and output blocks.	There are 10 layers 5 in encoder and 5 in decoder, it has two modes spatial and channel attention mode, both of these processes happen in encoder layer where max pooling happens, and in the decoder layer only spatial attention block exists.	The SCAU-Net has added the basic U-Net architecture with the spatial and channel attention networks, in this model the input process starts with the encoding of the input blocks(64) with max pooling and then it increases till block(512), then channel attention layer performs the Unpooling and decoding starts.	The up-sampling process consists of 5 initial layers which further divided into different layers, the up-sampling process performs max-pooling, similarly the down-sampling process consists of 5 layers and performs un-pooling process.	1. The major strength of this architecture is that it uses the channel and attention mode for the segmentation of the input images it uses the blocks with 64 then 128 and then 512 pixels then decode them to the output labels with the max pooling convolution of (1x1,C) matrix.	It is a model which is unlike other models give the results of 1% improved in the dice score and 1.5% in the jaccard score. The use of the spatial and channel attention gives much precise results with the dice score of 0.9100 and jaccard score of 0.8381.	The main disadvantage of this model is that the spatial and channel attention blocks are in both encoding and decoding layers it make training process slow and sometimes can give overfitted results.

Author and Year	Architectures	Encoders/decoders number	Layers added	Added architectures	Up/down-sampling numbers	The major difference	Strengths	Weakness
Simon A. A. Kohl *et al.* 2018 [14]	Probabilistic U-Net for segmentation of ambiguous images.	It used 5 VCG-16 encoders and decoder, it uses the 3 U-Net architectures and uses the applications of the latent space.	There are total 5 layers in the probabilistic U-Net and there is another layer of prior net connected with the U-Net and posterior net which is connected to latent space and ground truth images.	The probabilistic U-Net architecture is combination of the U-Net architecture, prior net and posterior net which use the application of the latent space and cross entropy which converts the ground truth images to segmentation label.	The probabilistic U-Net contains 1x1convolutions up-sampling label and its prior net use the application of cross entropy and in the down-sampling and batch-normalization in the latent space and prior net.	1. The most basic difference in the probabilistic U-Net than the others is that it uses the application of the latent space of posterior net and cross entropy of prior net and it makes it work more precisely than the basic architecture.	The advantage of this model is that it can work more precisely on the images which are ambiguous and gives the 1.5 % accurate result than the basic architecture.	The major weakness of this model is, that it can do the segmentation of the ambiguous and noisy images, but if the image is lesser noisy the result will be overfitted.
Xuebin Qin *et al.* (2020)[15]	U^2 Net deeper U-Net architecture	In the model there are 6 encoders /decoders networks used same as basic U-Net architecture in an exponential way and they are added with the residual U-Block(RSU) with SOD data.	There are 6 layers in the U^2 Net architecture, each layer consists of two U-Net architectures which work as encoder /decoder pair in the architectures which are added with the RSU block.	There are total of 6 architectures added in the form of the layers and used for up/down-sampling of the input images.	There are decreasing number of downsamples which is down-samplesx1/2 which means it goes decreasing after each process while the increasing number of up-samples which up-samples x 2 which means it increases in a multiplicative way.	The major strength of this model is that it uses the U-Net architecture with the Residual U Block and the U blocks are the elements of the encoding and decoding process, it does the segmentation at each layer and gives output at each until the last segmented image produces.	The strength of this model is that it gives results after every layer and gives the output and combine them to give the ultimate segmented image and hence it can give the precise result than the others.	The main disadvantage with this model is if one image isn't processed correctly then the same error will appear in all the output layers and it can effect on the ultimate results.

Author and Year	Architectures	Encoders/decoders number	Layers added	Added architectures	Up/down-sampling numbers	The major difference	Strengths	Weakness
Huimin Huang *et al.* (2020)[16]	U-Net3+ architecture a full scale architecture.	In the model there are combination of U-Net and U-Net++ including the encoder /decoder same as Unet++ models.	In the Unet++ model there is dense and nested connections layer, while in the basic U-net there is a plain skip connection layer, in the U-Net3+, the model consists of full scale skip connection layers.	There are 3 architecture added in the model which made it the combination of U-Net and U-Net++, the three architectures are U-Net architecture, U- Net++, and nested U-Net architectures.	The up-sampling max-pooling process decreases sequentially and the process completes by 64, 3x3 convolution in the down-sampling process the un-pooling process 1x1 convolution.	The major strength of this model is that unlike the other nested and densely connected model it is fully skip connected model which is combination of basic and nested U-Net architecture. It makes this model more adaptive and scalable.	The strength of this model is that it can give results more precisely than the Unet++ with dice score of 0.9552 while UNet++ has 0.9352.	The major weakness of this architecture is that it takes too much time train the model and thus it needs much amount GPU and a powerful processor for the completion of the process.

3.Our observations on the architecture based on U-Net

The basic U-Net architecture[1] works with the convolution neural network and it has some drawbacks like training process takes too much time, apart from that the U-Net architecture [3] which is used for image segmentation requires almost 12 GB of GPU and relatively high GPU memory footprint for higher resolution images. The architecture initially was designed to overcome the ad hoc choice of 'rolling window' / patches hyper-parameters.

To have decent results the size of U-NET that you will be using must be comparable with the size of features and its surroundings (i.e. you should include some context for the model to work properly). In the basic architectures there are few errors and performance drawbacks to overcome this, the first modified model was introduced by Debesh *et al. 2018* [6], this architecture was the Double U-Net [6].

The limitations of the double U-Net architecture overcame in the U-Net++ [13], in the basic architecture there are some limitations first one is their optimal depth is apriori unknown, requiring extensive architecture search or inefficient ensemble of models of varying depths and second is their skip connections impose an unnecessarily restrictive fusion scheme, forcing aggregation only at the same-scale feature maps of the encoder and decoder sub-networks. To overcome such drawbacks the model U-Net++ [13] is proposed by Zongwei *et al.* [13]and they have proposed the method called nested U-Net architecture where they have applied 'n' numbers of U-net added with each other in a feed forwarded network.

The disadvantages of this architecture is that it also works on the overlap tile strategy using 1X1 up sampling convolution and 4x4 down sampling convolutions so if the inserted training images are distorted then it can affect the whole segmentation results.

To improve the tile overlap strategy for the image segmentation the new model which is also a modification in the basic U-Net architecture is given by Jeya Maria Jose Valanarasu *et al. 2020* [5] and the model proposed by them is called the KiU-Net [5], they gave two architectures for the same KiU-Net[5] and KiU-Net 3d the KiU-Net is proposed to overcome the performance drop of the basic U-Net[3] and U-Net++[13] and to remove the noise from the segmented images, the KiU-Net [5] is an improved model with higher precision than U-Net++ and U-Net architecture. The precision of the KiU-Net is 0.8943 while the precision of the U-Net is 0.8537 and U-Net++ is 0.8624.

There are many enhanced model of basic architecture proposed by many researchers there is another model proposed by Peng Zhao *et al. 2020* [14] and the proposed by them was Spatial-Channel attention U-Net (SCAU-Net) architecture [14], this architecture use the application of deep learning and like basic architecture it also contains the encoder and decoder. The SCAU-Net has two modes these are spatial and channel attention modes. The SCAU-Net which has included both spatial and channel attention modes have higher precision and dice score than basic architecture, double U-Net [6] and U-Net++ [13], the SCAU-Net [14] has dice score of 0.9063 while basic architecture and others have the dice scores of 0.8952 and 0.8963.

Hence from the above observations, we can see that U-Net architecture is a powerful and highly useful architecture which is used for image segmentation applications, there are many updated models given by many researchers till now, there are other updated models too which are based on the U-Net and they are improved models of the basic U-Net architecture like V-Net architecture [26] proposed by Fausto *et al.* (2016) which is using application of CNN and deep learning for volumetric biomedical image segmentation.

The most recent proposed models of the basic architectures in 2020 are U^2-Net architecture [27] given by Xuebin Qin*et al.*(2020)and U-Net3+ architecture[28] given by Huimin Huang *et al.* (2020).The first one is using nested architecture with application of deep learning and is able to capture more contextual information from different scales because of the mixture of receptive fields of different sizes in the proposed ReSidual U-blocks (RSU) and in the second one U-Net 3+ architecture [28] is another improvement of basic architecture and U-Net++ [13] and takes advantage of full-scale skip connections and deep supervisions.

The full-scale skip connections incorporate low-level details with high-level semantics from feature maps in different scales;

while the deep supervision learns hierarchical representations from the full-scale aggregated feature maps.

Here we can see that the models given by the various researchers have some common aspects these are:

- These models are trained and evaluated on the same datasets, these are LIDC-IDRI, BraTs 2015, 17, 18 and 2020, and MICCAI-2015 and 2018 datasets.

- All the models are trained to enhance the techniques and methodology of basic U-Net Architecture[1] and to add some learning parameters with that, for example in the Double U-Net[6], the first U-Net is trained already by VCG-19 encoder while other encoder is trained from scratch.

- The models have used the applications of convolution network and almost all of them has applied the learning using convolution network, the basic U-Net architecture use the convolutional networks for the training and hence all the architectures based on U-Net are using the applications of convolutional networks for example: BCDU-Net [7], which uses the application of densely connected convolutions.

- All the proposed models require a high processing GPU system with the size of almost 12 GB for the processing and segmentation of the images which makes training process slow.

- All the proposed models consist of the encoder and decoder system and works on plug-play methodology, also they consists of pooling, max filters and activation functions.

- All of the given proposed models has added the basic U-Net architecture with the other training parameters and models, some of them has added more than one U-Net architectures together and made new model example: U-Net++[11] which has given nested U-Net architecture and Double U-Net[6] which has added two U-Net architectures together.

So, these are some of the common points in the given models which are given by many researchers and all the models work on the segmentation of the images. These models are proposed to overcome the drawbacks of the basic U-Net architectures [1], for example the basic architecture fails when the image contains too much noise or the image is too blur to go for segmentation. While the other models are used for the segmentation of noisy images and can do the segmentation better than the basic architecture.

Hence the U-Net architecture is highly recommended, scalable and improvable architecture which can be modified further and can be used for the segmentation of advanced biomedical images and can be used for the segmentation of the ambiguous images.

Conclusion

The U-Net architecture is an efficient network which takes the help of convolutional neural networks and there are many architectures whose base is the U-Net architecture and they work much efficiently and effectively on the images. But there are images which are complex and ambiguous that cannot be segmented easily by the help of the basic U-Net architecture and for that there are other novel architectures which are based on the U-Net architectures that are similar to basic architectures containing encoder, decoder, contracting and expanding path.

Some architectures use double U-Net networks, some of them use nested U-Net architecture while others are using modes in the network like spatial and channel attention mode. From the above tables we can simply classify which architecture is more efficient and which is less effective but in a nutshell, all the proposed architectures which are based on the novel U-Net architecture are working than the basic U-Net architecture with the help of their dice score and results on the datasets we can say that all the proposed architectures are better than the basic U-Net architecture.

Shiv Hari Tewari
M.Tech. in CS and Engg.

Chapter 9

Adapting The Change

Technological advancement has impacted most industries, creating an ever faster pace of change, as our digital environment continues to develop at a rapid pace, it is not possible for all leaders at all levels to understand and master the myriad new technologies that may impact their organizations and markets. Research demonstrated that leaders felt the need to understand the impact of digitisation on their industry in order to feel confident to hand over accountability for understanding and mastering niche technologies to digital and technical experts. Demands made of leaders are more complex and fluid in what is now a constantly fluctuating environment. Working across geographical, functional and cultural boundaries presents today's leaders with greater challenges than ever before and required continuous improvements in the quality of leadership.

How do we know that change is accelerating? There is, after all, no absolute way to measure change. In the awesome complexity of the universe, even within any given society, a virtually infinite number of streams of change occur simultaneously. All "things" from the tiniest virus to the greatest galaxy are, in reality, not things at all, but processes.

There is no static point, no nirvana-like un-change, against which to measure change. Change is, therefore, necessarily relative. It is also uneven. If all processes occurred at the same speed, or even if they accelerated or decelerated in unison, it would be impossible to observe change. The future, however, invades the present at differing speeds. Thus it becomes possible to compare the speed of different processes as they unfold.

We know, for example, that compared with the biological evolution of the species, cultural and social evolution is extremely rapid. We know that some societies transform themselves technologically or economically more rapidly than others. We also know that different sectors within the same society exhibit different rates of change the disparity that William Ogburn labeled "cultural lag." It is precisely the unevenness of change that makes it measurable. We need, however, a yardstick that makes it possible to compare highly diverse processes, and this yardstick is time. Without time, change has no meaning. And without change, time would stop. Time can be conceived as the intervals during which events occur.

Just as money permits us to place a value on both apples and oranges, time permits us to compare unlike processes. When we say that it takes three years to build a dam, we are really saying it takes three times as long as it takes the earth to circle the sun or 31,000,000 times as long as it takes to sharpen a pencil. Time is the currency of exchange that makes it possible to compare the rates at which very different processes play themselves out.

Given the unevenness of change and armed with this yardstick, we still face exhausting difficulties in measuring change. When we speak of the rate of change, we refer to the number of events crowded into an arbitrarily fixed interval of time. Thus we need to define the "events." We need to select our intervals with precision. We need to be careful about the conclusions we draw from the differences we observe. Moreover, in the measurement of change, we are today far more advanced with respect to physical processes than social processes. We know far better, for example, how to measure the rate at which blood flows through the body than the rate at which a rumor flows through society.

Even with all these qualifications, however, there is widespread agreement, reaching from historians and archaeologists all across the spectrum to scientists, sociologists, economists and psychologists, that, many social processes are speeding up strikingly, even spectacularly.

Painting with the broadest of brush strokes, biologist Julian Huxley informs us that "The tempo of human evolution during recorded history is at least 100,000 times as rapid as that of pre-human evolution." Inventions or improvements of a magnitude that took perhaps 50,000 years to accomplish during the early Paleolithic era were, he says, "run through in a mere millennium toward its close; and with the advent of settled civilization, the unit of change soon became reduced to the century." The rate of change, accelerating throughout the past
5000 years, has become, in his words, "particularly noticeable during the past 300 years."

C. P. Snow, the novelist and scientist, also comments on the new visibility of change. "Until this century" he writes, social change was "so slow, that it would pass unnoticed in one person's lifetime. That is no longer so. The rate of change has increased so much that our imagination can't keep up." Indeed, says social psychologist Warren Bennis, the throttle has been pushed so far forward in recent years that "No exaggeration, no hyperbole, no outrage can realistically describe the extent and pace of change. In fact, only the exaggerations appear to be true."

What changes justify such super-charged language? Let us look at a few change in the process by which man forms cities, for example. We are now undergoing the most extensive and rapid urbanization the world has ever seen. In 1850 only four cities on the face of the earth had a population of 1,000,000 or more. By 1900 the number had increased to nineteen. But by 1960, there were 141, and today world urban population is rocketing upward at a rate of 6.5 percent per year, according to Edgar de Vries and J. P. Thysse of the Institute of Social Science in The Hague. This single stark statistic means a doubling of the earth's urban population within eleven years.

One way to grasp the meaning of change on so phenomenal a scale is to imagine what would happen if all existing cities, instead of expanding, retained their present size. If this were so, in order to accommodate the new urban millions we would have to build a duplicate city for each of the hundreds that already dot the globe. A new Tokyo, a new Hamburg, a new Rome and Rangoon and all within eleven years. (This explains why French urban planners are sketching subterranean cities stores, museums, warehouses and factories to be built under the earth, and why a Japanese architect has blueprinted a city to be built on stilts out over the ocean.)

The same accelerative tendency is instantly apparent in man's consumption of energy. Dr. Homi Bhabha, the late Indian atomic scientist who chaired the first International Conference on the Peaceful Uses of Atomic Energy, once analyzed this trend. "To illustrate," he said, "let us use the letter 'Q' to stand for the energy derived from burning some 33,000 million tons of coal. In the eighteen and one half centuries after Christ, the total energy consumed averaged less than one half Q per century. But by 1850, the rate had risen to one Q per century. Today, the rate is about ten Q per century." This means, roughly speaking, that half of all the energy consumed by man in the past 2,000 years has been consumed in the last one hundred.

Also dramatically evident is the acceleration of economic growth in the nations now racing toward super-industrialism. Despite the fact that they start from a large industrial base, the annual percentage increases in production in these countries are formidable. And the rate of increase is itself increasing.

In France, for example, in the twenty-nine years between 1910 and the outbreak of the second world war, industrial production rose only 5 percent. Yet between 1948 and 1965, in only seventeen years, it increased by roughly 220 percent. Today growth rates of from 5 to 10 percent per year are not uncommon among the most industrialized nations. There are ups and downs, of course. But the direction of change has been unmistakable.

Thus for the twenty-one countries belonging to the Organization for Economic Cooperation and Development by and large, the "have" nations the average annual rate of increase in gross national product in the years 1960-1968 ran between 4.5 and 5.0 percent.

The United States grew at a rate of 4.5 percent, and Japan led the rest with annual increases averaging 9.8 percent. What such numbers imply is nothing less revolutionary than a doubling of the total output of goods and services in the advanced societies about every fifteen years and the doubling times are shrinking.

This means, generally speaking, that the child reaching teen age in any of these societies is literally surrounded by twice as much of everything newly man made as his parents were at the time he was an infant. It means that by the time today's teenager reaches age thirty, perhaps earlier, a second doubling will have occurred. Within a seventy-year lifetime, perhaps five such doublings will take place meaning, since the increases are compounded, that by the time the individual reaches old age the society around him will be producing thirty-two times as much as when he was born.

Such changes in the ratio between old and new have, as we shall show, an electric impact on the habits, beliefs, and self-image of millions. Never in previous history has this ratio been transformed so radically in so brief a flick of time.

Behind such prodigious economic facts lies that great, growling engine of change technology. This is not to say that technology is the only source of change in society. Social upheavals can be touched off by a change in the chemical composition of the atmosphere, by alterations in climate, by changes in fertility, and many other factors. Yet technology is indisputably a major force behind the accelerative thrust.

To most people, the term technology conjures up images of smoky steel mills or clanking machines. Perhaps the classic symbol of technology is still the assembly line created by Henry Ford half a century ago and made into a potent social icon by Charlie Chaplin in Modern Times. This symbol, however, has always been inadequate, indeed, misleading, for technology has always been more than factories and machines.

The invention of the horse collar in the middle ages led to major changes in agricultural methods and was as much a technological advance as the invention of the Bessemer furnace centuries later. Moreover, technology includes techniques, as well as the machines that may or may not be necessary to apply them. It includes ways to make chemical reactions occur, ways to breed fish, plant forests, light theaters, count votes or teach history.

The old symbols of technology are even more misleading today, when the most advanced technological processes are carried out far from assembly lines or open hearths. Indeed, in electronics, in space technology, in most of the new industries, relative silence and clean surroundings are characteristic even sometimes essential. And the assembly line the organization of armies of men to carry out simple repetitive functions is an anachronism. It is time for our symbols of technology to change to catch up with the quickening changes in technology, itself.

This acceleration is frequently dramatized by a thumbnail account of the progress in transportation. It has been pointed out, for example, that in 6000 B.C. the fastest transportation available to man over long distances was the camel caravan, averaging eight miles per hour. It was not until about 1600 B.C. when the chariot was invented that the maximum speed was raised to roughly twenty miles per hour.

So impressive was this invention, so difficult was it to exceed this speed limit, that nearly 3,500 years later, when the first mail coach began operating in England in 1784, it averaged a mere ten mph. The first steam locomotive, introduced in 1825, could muster a top speed of only thirteen mph, and the great sailing ships of the time labored along at less than half that speed. It was probably not until the 1880's that man, with the help of a more advanced steam locomotive, managed to reach a speed of one hundred mph. It took the human race millions of years to attain that record.

It took only fifty-eight years, however, to quadruple the limit, so that by 1938 airborne man was cracking the 400-mph line. It took a mere twenty-year flick of time to double the limit again. And by the 1960's rocket planes approached speeds of 4000 mph, and men in space capsules were circling the earth at 18,000 mph. Plotted on a graph, the line representing progress in the past generation would leap vertically off the page. Whether we examine distances traveled, altitudes reached, minerals mined, or explosive power harnessed, the same accelerative trend is obvious. The pattern, here and in a thousand other statistical series, is absolutely clear and unmistakable. Millennia or centuries go by, and then, in our own times, a sudden bursting of the limits, a fantastic spurt forward.

The reason for this is that technology feeds on itself. Technology makes more technology possible, as we can see if we look for a moment at the process of innovation. Technological innovation consists of three stages, linked together into a self-reinforcing cycle. First, there is the creative, feasible idea. Second, its practical application. Third, its diffusion through society.The process is completed, the loop closed, when the diffusion of technology embodying the new idea, in turn, helps generate new creative ideas. Today there is evidence that the time between each of the steps in this cycle has been shortened.

Thus it is not merely true, as frequently noted, that 90 percent of all the scientists who ever lived are now alive, and that new scientific discoveries are being made every day. These new ideas are put to work much more quickly than ever before. The time between original concept and practical use has been radically reduced. This is a striking difference between ourselves and our ancestors. Appollonius of Perga discovered conic sections, but it was 2000 years before they were applied to engineering problems. It was literally centuries between the time Paracelsus discovered that ether could be used as an anaesthetic and the time it began to be used for that purpose.

Even in more recent times the same pattern of delay was present. In 1836 a machine was invented that mowed, threshed, tied straw into sheaves and poured grain into sacks. This machine was itself based on technology at least twenty years old at the time. Yet it was not until a century later, in the 1930's, that such a combine was actually marketed. The first English patent for a typewriter was issued in 1714. But a century and a half elapsed before typewriters became commercially available. A full century passed between the time Nicholas Appert discovered how to can food and the time canning became important in the food industry.

Today such delays between idea and application are almost unthinkable. It is not that we are more eager or less lazy than our ancestors, but we have, with the passage of time, invented all sorts of social devices to hasten the process. Thus we find that the time between the first and second stages of the innovative cycle between idea and application has been cut radically. Frank Lynn, for example, in studying twenty major innovations, such as frozen food, antibiotics, integrated circuits and synthetic leather, found that since the beginning of this century more than sixty percent has been slashed from the average time needed for a major scientific discovery to be translated into a useful technological form. Today a vast and growing research and development industry is consciously working to reduce the lag still further. But if it takes less time to bring a new idea to the marketplace, it also takes less time for it to sweep through the society.

Thus the interval between the second and third stages of the cycle between application and diffusion has likewise been sliced, and the pace of diffusion is rising with astonishing speed. This is borne out by the history of several familiar household appliances. Robert B. Young at the Stanford Research Institute has studied the span of time between the first commercial appearance of a new electrical appliance and the time the industry manufacturing it reaches peak production of the item.

Young found that for a group of appliances introduced in the United States before 1920 including the vacuum cleaner, the electric range, and the refrigerator the average span between introduction and peak production was thirty-four years. But for a group that appeared in the 1939-1959 period including the electric frying pan, television, and washer dryer combination the span was only eight years. The lag had shrunk by more than 76 percent. "The post-war group," Young declared, "demonstrated vividly the rapidly accelerating nature of the modern cycle."

The stepped-up pace of invention, exploitation, and diffusion, in turn, accelerates the whole cycle still further. For new machines or techniques are not merely a product, but a source, of fresh creative ideas.

Each new machine or technique, in a sense, changes all existing machines and techniques, by permitting us to put them together into new combinations. The number of possible combinations rises exponentially as the number of new machines or techniques rises arithmetically. Indeed, each new combination may, itself, be regarded as a new super machine. The computer, for example, made possible a sophisticated space effort. Linked with sensing devices, communications equipment, and power sources, the computer became part of a configuration that in aggregate forms a single new super-machine a machine for reaching into and probing outer space. But for machines or techniques to be combined in new ways, they have to be altered, adapted, refined or otherwise changed. So that the very effort to integrate machines into super-machines compels us to make still further technological innovations.

It is vital to understand, moreover, that technological innovation does not merely combine and recombine machines and techniques. Important new machines do more than suggest or compel changes in other machines they suggest novel solutions to social, philosophical, even personal problems. They alter man's total intellectual environment the way he thinks and looks at the world.

We all learn from our environment, scanning it constantly though perhaps unconsciously for models to emulate. These models are not only other people. They are, increasingly, machines. By their presence, we are subtly conditioned to think along certain lines. It has been observed, for example, that the clock came along before the Newtonian image of the world as a great clock-like mechanism, a philosophical notion that has had the utmost impact on man's intellectual development. Implied in this image of the cosmos as a great clock were ideas about cause and effect and about the importance of external, as against internal, stimuli, that shape the everyday behavior of all of us today.

The clock also affected our conception of time so that the idea that a day is divided into twenty-four equal segments of sixty minutes each has become almost literally a part of us.Recently, the computer has touched off a storm of fresh ideas about man as an interacting part of larger systems, about his physiology, the way he learns, the way he remembers, the way he makes decisions. Virtually every intellectual discipline from political science to family psychology has been hit by a wave of imaginative hypotheses triggered by the invention and diffusion of the computer and its full impact has not yet struck. And so the innovative cycle, feeding on itself, speeds up.

If technology, however, is to be regarded as a great engine, a mighty accelerator, then knowledge must be regarded as its fuel. And we thus come to the crux of the accelerative process in society, for the engine is being fed a richer and richer fuel every day.

Knowledge As Fuel

The rate at which man has been storing up useful knowledge about himself and the universe has been spiraling upward for 10,000 years. The rate took a sharp upward leap with the invention of writing, but even so it remained painfully slow over centuries of time. The next great leap forward in knowledge acquisition did not occur until the invention of movable type in the fifteenth century by Gutenberg and others. Prior to 1500, by the most optimistic estimates, Europe was producing books at a rate of 1000 titles per year.

This means, give or take a bit, that it would take a full century to produce a library of 100,000 titles. By 1950, four and a half centuries later, the rate had accelerated so sharply that Europe was producing 120,000 titles a year. What once took a century now took only ten months. By 1960, a single decade later, the rate had made another significant jump, so that a century's work could be completed in seven and a half months. And, by the mid-sixties, the output of books on a world scale, Europe included, approached the prodigious figure of 1000 titles per day.One can hardly argue that every book is a net gain for the advancement of knowledge.

Nevertheless, we find that the accelerative curve in book publication does, in fact, crudely parallel the rate at which man discovered new knowledge. For example, prior to Gutenberg only 11 chemical elements were known. Antimony, the 12th, was discovered at about the time he was working on his invention. It was fully 200 years since the 11th, arsenic, had been discovered. Had the same rate of discovery continued, we would by now have added only two or three additional elements to the periodic table since Gutenberg. Instead, in the 450 years after his time, some seventy additional elements were discovered. And since 1900 we have been isolating the remaining elements not at a rate of one every two centuries, but of one every three years.

Furthermore, there is reason to believe that the rate is still rising sharply. Today, for example, the number of scientific journals and articles is doubling, like industrial production in the advanced countries, about every fifteen years, and according to biochemist Philip Siekevitz, "what has been learned in the last three decades about the nature of living beings dwarfs in extent of knowledge any comparable period of scientific discovery in the history of mankind." Today the United States government alone generates 100,000 reports each year, plus 450,000 articles, books and papers. On a worldwide basis, scientific and technical literature mounts at a rate of some 60,000,000 pages a year.

The computer burst upon the scene around 1950. With its unprecedented power for analysis and dissemination of extremely varied kinds of data in unbelievable quantities and at mind-staggering speeds, it has become a major force behind the latest acceleration in knowledge-acquisition. Combined with other increasingly powerful analytical tools for observing the invisible universe around us, it has raised the rate of knowledge-acquisition to dumbfounding speeds. Francis Bacon told us that "Knowledge ... is power." This can now be translated into contemporary terms. In our social setting, "Knowledge is change" and accelerating knowledge-acquisition, fueling the great engine of technology, means accelerating change.

The Concept of Transience

Much of our theorizing about social and psychological change presents a valid picture of man in relatively static societies but a distorted and incomplete picture of the truly contemporary man. It misses a critical difference between the men of the past or present and the men of the future. This difference is summed up in the word "transience."

The concept of transience provides a long-missing link between sociological theories of change and the psychology of individual human beings. Integrating both, it permits us to analyze the problems of high-speed change in a new way. And, as we shall see, it gives us a method crude but powerful to measure inferentially the rate of situation flow.

Transience is the new "temporariness" in everyday life. It results in a mood, a feeling of impermanence. Philosophers and theologians, of course, have always been aware that man is ephemeral. In this grand sense, transience has always been a part of life. But today the feeling of impermanence is more acute and intimate. Thus Edward Albee's character, Jerry, in The Zoo Story, characterizes himself as a "permanent transient." And critic Harold Clurman, commenting on Albee, writes: "None of us occupy abodes of safety true homes. We are all the same 'people in all the rooming houses everywhere,' desperately and savagely trying to effect soul-satisfying connections with our neighbors." We are, in fact, all citizens of the Age of Transience.

It is, however, not only our relationships with people that seem increasingly fragile or impermanent. If we divide up man's experience of the world outside himself, we can identify certain classes of relationships. Thus, in addition to his links with other people, we may speak of the individual's relationship with things. We can single out for examination his relationships with places. We can analyze his ties to the institutional or organizational environment around him. We can even study his relationship to certain ideas or to the information flow in society.

These five relationships plus time form the fabric of social experience. This is why, as suggested earlier, things, places, people, organizations and ideas are the basic components of all situations. It is the individual's distinctive relationship to each of these components that structures the situation. And it is precisely these relationships that, as acceleration occurs in society, become foreshortened, telescoped in time. Relationships that once endured for long spans of time now have shorter life expectancies. It is this abbreviation, this compression, that gives rise to the almost tangible feeling that we live, rootless and uncertain, among shifting dunes.

Transience, indeed, can be defined quite specifically in terms of the rate at which our relationships turn over. While it may be difficult to prove that situations, as such, take less time to pass through our experience than before, it is possible to break them down into their components, and to measure the rate at which these components move into and out of our lives to measure, in other words, the duration of relationships.

It will help us understand the concept of transience if we think in terms of the idea of "turnover." In a grocery store, for example, milk turns over more rapidly than, say, canned asparagus. It is sold and replaced more rapidly. The "through-put" is faster. The alert businessman knows the turnover rate for each of the items he sells, and the general rate for the entire store. He knows, in fact, that his turnover rate is a key indicator of the health of the enterprise.

We can, by analogy, think of transience as the rate of turnover of the different kinds of relationships in an individual's life. Moreover, each of us can be characterized in terms of this rate. For some, life is marked by a much slower rate of turnover than for others. The people of the past and present lead lives of relatively "low transience" their relationships tend to be long lasting. But the people of the future live in a condition of "high transience" a condition in which the duration of relationships is cut short, the through-put of relationships extremely rapid. In their lives, things, places, people, ideas, and organizational structures all get "used up" more quickly.

This affects immensely the way they experience reality, their sense of commitment, and their ability or inability to cope. It is this fast through-put, combined with increasing newness and complexity in the environment, that strains the capacity to adapt and creates the danger of future shock. If we can show that our relationships with the outer world are, in fact, growing more and more transient, we have powerful evidence for the assumption that the flow of situations is speeding up. And we have an incisive new way of looking at ourselves and others. Let us, therefore, explore life in a high transience society.

The FAD Machine

Fast-shifting preferences, flowing out of and interacting with high-speed technological change, not only lead to frequent changes in the popularity of products and brands, but also shorten the life cycle of products. Automation expert John Diebold never wearies of pointing out to businessmen that they must begin to think in terms of shorter life spans for their goods. Smith Brothers' Cough Drops, Calumet Baking Soda and Ivory Soap, have become American institutions by virtue of their long reign in the market place.

In the days ahead, he suggests, few products will enjoy such longevity. Every consumer has had the experience of going to the supermarket or department store to replace some item, only to find that he cannot locate the same brand or product. In 1966 some 7000 new products turned up in American supermarkets. Fully 55 percent of all the items now sold there did not exist ten years ago. And of the products available then, 42 percent have faded away altogether. Each year the process repeats itself in more extreme form. Thus 1968 saw 9,500 new items in the consumer packaged-goods field alone, with only one in five meeting its sales target.

A silent but rapid attrition kills off the old, and new products sweep in like a ride. "Products that used to sell for twenty-five years," writes economist Robert Theobald, "now often count on no more than five. In the volatile pharmaceutical and electronic fields the period is often as short as six months." As the pace of change accelerates further, corporations may create new products knowing full well that they will remain on the market for only a matter of a few weeks.

Here, too, the present already provides us with a foretaste of the future. It lies in an unexpected quarter: the fads now sweeping over the high technology societies in wave after wave. In the past few years alone, in the United States, Western Europe and Japan, we have witnessed the sudden rise or collapse in popularity of "Bardot hairdos," the "Cleopatra look," James Bond, and Batman, not to speak of Tiffany lampshades, Super-Balls, iron crosses, pop sunglasses, badges and buttons with protest slogans or pornographic jokes, posters of Allen Ginsberg or Humphrey Bogart, false eyelashes, and innumerable other gimcracks and oddities that reflect are tuned into the rapidly changing pop culture.

Backed by mass media promotion and sophisticated marketing, such fads now explode on the scene virtually overnight and vanish just as quickly. Sophisticates in the fad business prepare in advance for shorter and shorter product life cycles. Thus, there is in San Gabriel, California, a company entitled, with a kind of cornball relish, Wham-O Manufacturing Company. Wham-O specializes in fad products, having introduced the hula hoop in the fifties and the so-called Super-Ball more recently.

The latter a high-bouncing rubber ball quickly became so popular with adults as well as children that astonished visitors saw several of them bouncing merrily on the floor of the Pacific Coast Stock Exchange. Wall Street executives gave them away to friends and one high broadcasting official complained that "All our executives are out in the halls with their Super-Balls." Wham-O, and other companies like it, however, are not disconcerted when sudden death overtakes their product; they anticipate it. They are specialists in the design and manufacture of "temporary" products.

The fact that fads are generated artificially, to a large extent, merely underscores their significance. Even engineered fads are not new to history. But never before have they come fleeting across the consciousness in such rapid-fire profusion, and never has there been such smooth coordination between those who originate the fad, mass media eager to popularize it, and companies geared for its instantaneous exploitation. A well-oiled machinery for the creation and diffusion of fads is now an entrenched part of the modern economy. Its methods will increasingly be adopted by others as they recognize the inevitability of the ever-shorter product cycle. The line between "fad" and ordinary product will progressively blur. We are moving swiftly into the era of the temporary product, made by temporary methods, to serve temporary needs.

The turnover of things in our lives thus grows even more frenetic. We face a rising flood of throw-away items, impermanent architecture, mobile and modular products, rented goods and commodities designed for almost instant death. From all these directions, strong pressures converge toward the same end: the inescapable ephemeralization of the man-thing relationship.The foreshortening of our ties with the physical environment, the stepped-up turnover of things, however, is only a small part of a much larger context. Let us, therefore, press ahead in our exploration of life in high transience society.

Information

In a society in which instant food, instant education and even instant cities are everyday phenomena, no product is more swiftly fabricated or more ruthlessly destroyed than the instant celebrity. Nations advancing toward super-industrialism sharply step up their output of these "psycho-economic" products. Instant celebrities burst upon the consciousness of millions like an image-bomb which is exactly what they are. thousands of "personalities" parade across the stage of contemporary history. Real people, magnified and projected by the mass media, they are stored as images in the minds of millions of people who have never met them, never spoken to them, never seen them "in person." They take on a reality almost as (and sometimes even more) intense than that of many people with whom we do have "in-person" relationships.

We form relationships with these "vicarious people," just as we do with friends, neighbors and colleagues. And just as the through-put of real, in-person people in our lives is increasing, and the duration of our average relationship with them decreasing, the same is true of our ties with the vicarious people who populate our minds. Their rate of flow-through is influenced by the real rate of change in the world. Thus, in politics, for example, we find that the British prime minister ship has been turning over since 1922 at a rate some 13 percent faster than in the base period 1721-1922. In sports, the heavyweight boxing championship now changes hands twice as fast as it did during our father's youth.

Events, moving faster, constantly throw new personalities into the charmed circle of celebrity-hood, and old images in the mind decay to make way for the new.The same might be said for the fictional characters spewed out from the pages of books,
from television screens, theaters, movies and magazines. No previous generation in history has had so many fictional characters flung at it.

Commenting on the mass media, historian Marshall Fishwick wryly declares: "We may not even get used to Super-Hero, Captain Nice and Mr. Terrific before they fly off our television screens forever." These vicarious people, both live and fictional, play a significant role in our lives, providing models for behavior, acting out for us various roles and situations from which we draw conclusions about our own lives. We deduce lessons from their activities, consciously or not. We learn from their triumphs and tribulations.

They make it possible for us to "try on" various roles or life styles without suffering the consequences that might attend such experiments in real life. The accelerated flow-through of vicarious people cannot but contribute to the instability of personality patterns among many real people who have difficulty in finding a suitable life style. These vicarious people, however, are not independent of one another.
They perform their roles in a vast, complexly organized "public drama" which is, in the words of sociologist Orrin Klapp, author of a fascinating book called Symbolic Leaders, largely a product of the new communications technology. This public drama, in which celebrities upstage and replace celebrities at an accelerating rate, has the effect, according to Klapp, of making leadership "more unstable than it would be otherwise. Contretemps, upsets, follies, contests, scandals, make a feast of entertainment or a spinning political roulette wheel. Fads come and go at a dizzying pace. A country like the United States has an open public drama, in which new faces appear daily, there is always a contest to steal the show, and almost anything can happen and often does." What we are observing, says Klapp, is a "rapid turnover of symbolic leaders."

This can be extended, however, into a far more powerful statement: what is happening is not merely a turnover of real people or even fictional characters, but a more rapid turnover of the images and image-structures in our brains. Our relationships with these images of reality, upon which we base our behavior, are growing, on average, more and more transient.

The entire knowledge system in society is undergoing violent upheaval. The very concepts and codes in terms of which we think are turning over at a furious and accelerating pace. We are increasing the rate at which we must form and forget our images of reality.

Dr. Prabhat Kumar Pandey & Deepak Bishnoi

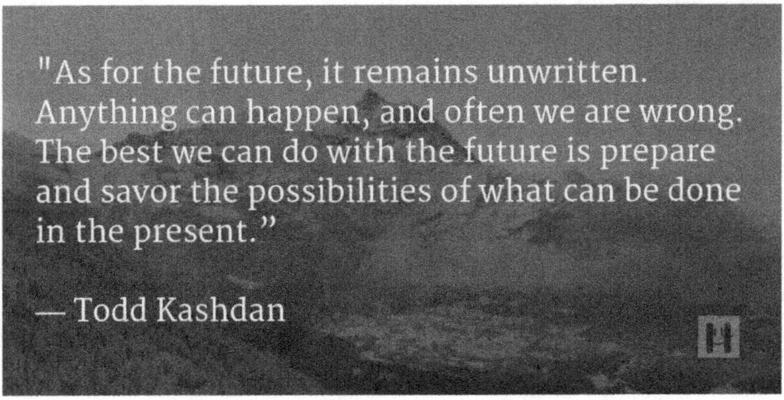
"As for the future, it remains unwritten. Anything can happen, and often we are wrong. The best we can do with the future is prepare and savor the possibilities of what can be done in the present."
— Todd Kashdan

Chapter 10

Survive With The Change

We are creating a new society. Not a changed society. Not an extended, larger-than-life version of our present society. But a new society. This simple premise has not yet begun to tincture our consciousness. Yet unless we understand this, we shall destroy ourselves in trying to cope with tomorrow.

A revolution shatters institutions and power relationships. This is precisely what is happening today in all the high-technology nations. Students in Berlin and New York, in Turin and Tokyo, capture their deans and chancellors, bring great clanking education factories to a grinding halt, and even threaten to topple governments. Police stand aside in the ghettos of New York, Washington and Chicago as ancient property laws are openly violated. Sexual standards are overthrown. Great cities are paralyzed by strikes, power failures, riots.

International power alliances are shaken. Financial and political leaders secretly tremble not out of fear that communist (or capitalist) revolutionaries will oust them, but that the entire system is somehow flying out of control. These are indisputable signs of a sick social structure, a society that can no longer perform even its most basic functions in the accustomed ways.

It is a society caught in the agony of revolutionary change. In the 1920's and 1930's, communists used to speak of the "general crisis of capitalism." It is now clear that they were thinking small. What is occurring now is not a crisis of capitalism, but of industrial society itself, regardless of its political form. We are simultaneously experiencing a youth revolution, a sexual revolution, a racial revolution, a colonial revolution, an economic revolution, and the most rapid and deep-going technological revolution in history. We are living through the general crisis of industrialism.

In a word, we are in the midst of the super-industrial revolution. If failure to grasp this fact impairs one's ability to understand the present, it also leads otherwise intelligent men into total stupidity when they talk about the future. It encourages them to think in simple-minded straight lines. Seeing evidence of bureaucracy today, they naïvely assume there will be more bureaucracy tomorrow. Such linear projections characterize most of what is said or written about the future. And it causes us to worry about precisely the wrong things.

One needs imagination to confront a revolution. For revolution does not move in straight lines alone. It jerks, twists and backtracks. It arrives in the form of quantum jumps and dialectical reversals. Only by accepting the premise that we are racing toward a wholly new stage of eco-technological development—the super-industrial stage can we make sense of our era. Only by accepting the revolutionary premise can we free our imaginations to grapple with the future.

Revolution implies novelty. It sends a flood of newness into the lives of countless individuals, confronting them with unfamiliar institutions and first-time situations. Reaching deep into our personal lives, the enormous changes ahead will transform traditional family structures and sexual attitudes. They will smash conventional relationships between old and young. They will overthrow our values with respect to money and success. They will alter work, play and education beyond recognition. And they will do all this in a context of spectacular, elegant, yet frightening scientific advance.

If transience is the first key to understanding the new society, therefore, novelty is the second. The future will unfold as an unending succession of bizarre incidents, sensational discoveries, implausible conflicts, and wildly novel dilemmas. This means that many members of the super-industrial society will never "feel at home" in it. Like the voyager who takes up residence in an alien country, only to find, once adjusted, that he must move on to another, and yet another, we shall come to feel like "strangers in a strange land."

The super-industrial revolution can erase hunger, disease, ignorance and brutality. Moreover, despite the pessimistic prophecies of the straight-line thinkers, super-industrialism will not restrict man, will not crush him into bleak and painful uniformity. In contrast, it will radiate new opportunities for personal growth, adventure and delight. It will be vividly colorful and amazingly open to individuality. The problem is not whether man can survive regimentation and standardization.

The problem, as we shall see, is whether he can survive freedom. Yet for all this, man has never truly inhabited a novelty-filled environment before. Having to live at an accelerating pace is one thing when life situations are more or less familiar. Having to do so when faced by unfamiliar, strange or unprecedented situations is distinctly another. By unleashing the forces of novelty, we slam men up against the non routine, the unpredicted. And, by so doing, we escalate the problems of adaptation to a new and dangerous level. For transience and novelty are an explosive mix.

If all this seems doubtful, let us contemplate some of the novelties that lie in store for us. Combining rational intelligence with all the imagination we can command, let us project ourselves forcefully into the future. In doing so, let us not fear occasional error the imagination is only free when fear of error is temporarily laid aside. Moreover, in thinking about the future, it is better to err on the side of daring, than the side of caution.

One sees why the moment one begins listening to the men who are even now creating that future. Listen, as they describe some of the developments waiting to burst from their laboratories and factories.

Migration to the future

There are, however, important differences between the kind of people who are on the move in the United States and those caught up in the European migrations. In Europe most of the new mobility can be attributed to the continuing transition from agriculture to industry; from the past to the present, as it were. Only a small part is as yet associated with the transition from industrialism to super-industrialism. In the United States, by contrast, the continuing redistribution of population is no longer primarily caused by the decline of agricultural employment. It grows, instead, out of the spread of automation and the new way of life associated with super-industrial society, the way of life of the future.

This becomes plain if we look at who is doing the moving in the United States. It is true that some technologically backward and disadvantaged groups, such as urban Negroes, are characterized by high rates of geographical mobility, usually within the same neighborhood or county. But these groups form only a relatively small slice of the total population, and it
would be a serious mistake to assume that high rates of geographical mobility correlate only with poverty, unemployment or ignorance. In fact, we find that men with at least one year of college education (an ever increasing group) move more, and further, than those without.

Thus we find that the professional and technical populations are among the most mobile of all Americans. And we find an increasing number of affluent executives who move far and frequently. (It is a house joke among executives of the International Business Machine Corporation that IBM stands for "I've Been Moved.") In the emerging super-industrialism it is precisely these groups professional, technical and managerial who increase in both absolute number and as a proportion of the total work force. They also give the society its characteristic flavor, as the denim-clad factory worker did in the past.

Just as millions of poverty-stricken and unemployed rural workers are flowing from the agricultural past into the industrial present in Europe, so thousands of European scientists, engineers and technicians are flowing into the United States and Canada, the most super industrial of nations. In West Germany, Professor Rudolf Mossbauer, a Nobel prizewinner in physics, announces that he is thinking of migrating to America because of disagreements over administrative and budgetary policies at home. Europe's political ministers, worried over the "technology gap," have looked on helplessly as Westinghouse, Allied Chemical, Douglas Aircraft, General Dynamics and other major American corporations sent talent scouts to London or Stockholm to lure away everyone from astrophysicists to turbine engineers.

But there is a simultaneous "brain-drain" inside the United States, with thousands of scientists and engineers moving back and forth like particles in an atom. There are, in fact, well recognized patterns of movement. Two major streams, one from the North and the other from the South, both converge in California and the other Pacific Coast states, with a way station at Denver. Another major stream flows up from the South toward Chicago and Cambridge, Princeton and Long Island. A counter-stream carries men back to the space and electronics industries in Florida.

A typical young space engineer of my acquaintance quit his job with RCA at Princeton to go to work for General Electric. The house he had purchased only two years before was sold; his family moved into a rented house just outside Philadelphia, while a new one was built for them. They will move into this new house the fourth in about five years provided he is not transferred or offered a better job elsewhere. And all the time, California beckons. There is a less obvious geographical pattern to the movement of management men, but, if anything, the turnover is heavier.

A decade ago William Whyte, in The Organization Man, declared that "The man who leaves home is not the exception in American society but the key to it. Almost by definition, the organization man is a man who left home and kept on going." His characterization, correct then, is even truer today.

The Wall Street Journal refers to "corporate gypsies" in an article headlined "How Executive Family Adapts to Incessant Moving About Country." It describes the life of M. E. Jacobson, an executive with the Montgomery Ward retail chain. He and his wife, both forty-six at the time the story appeared, had moved twenty-eight times in twenty-six years of married life. "I almost feel like we're just camping," his wife tells her visitors. While their case is atypical, thousands like them move on the average of once every two years, and their numbers multiply. This is true not merely because corporate needs are constantly shifting, but also because top management regards frequent relocation of its potential successors as a necessary step in their training.

This moving of executives from house to house as if they were life-size chessmen on a continent-sized board has led one psychologist to propose facetiously a money-saving system called "The Modular Family." Under this scheme, the executive not only leaves his house behind, but his family as well. The company then finds him a matching family (personality characteristics carefully selected to duplicate those of the wife and children left behind) at the new site. Some other itinerant executive then "plugs into" the family left behind. No one appears to have taken the idea seriously yet. In addition to the large groups of professionals, technicians and executives who engage in a constant round of "musical homes," there are many other peculiarly mobile groupings in the society.

A large military establishment includes tens of thousands of families who, peacetime and wartime, move again and again. "I'm not decorating any more houses," snaps the wife of an army colonel with irony in her voice: "The curtains never fit from one house to the next and the rug is always the wrong size or color. From now on I'm decorating my car."

Tens of thousands of skilled construction workers add to the flow. On another level are the more than 750,000 students attending colleges away from their home state, plus the hundreds of thousands more who are away from home but still within their home state. For millions, and particularly for the "people of the future," home is where you find it.

You’ve heard about it all. Seen it all. Read it all. These technologies no longer hold any secrets for you. Hell, you've probably mentioned them over dinner or at work and have become the go-to person for questions about future innovations.Yet, technology is ever-changing, and this precious knowledge must be both managed and updated regularly. With this in mind, list of the top 10 future technologies that are not on the public’s radar as of today but are likely to make big waves in the future.

1. Femtosecond Projection Two-Photon Lithography

3D printing is a solution looking for a problem, and over the past decade, the technology has failed to find a target audience. On the one hand, 3D printers are still too expensive for the average Joe; on the other, they are not sophisticated and fast enough for large-scale manufacturing companies.

This may change over the next few years: researchers have developed a method that uses lasers to project millions of points simultaneously onto 3D-printing material, instead of using one point at a time. And because they're bad at branding, they called it Femtosecond projection TPL. To easily understand FP-TPL, simply imagine using a million heated needles to strategically melt a block of wax versus using a single needle. This means that incredibly tiny structures can be 3D-printed much much faster (a thousand times faster, give or take), while still ensuring a good quality of the build.

So far, teams working on the innovation seem focused on flexible electronics and micro-optics. However, quick discoveries around materials (both liquid and solid) have led researchers to think that they will be able to build small but imagination-baffling structures in the near future.

Once the quality can be controlled over large scales, one could easily imagine this technology being used for the creation of healthcare-related nanorobots, allowing for the treatment of a multitude of diseases on the molecular level.

2. LiFi

The name says it all: LiFi aims to use light to transmit information from point A to point B. The technology works by encoding digital data and turning LED bulbs on and off faster than humans can notice to transfer it. The light then travels to a photoreceptor, which can decode and translate the data to a more classic radio frequency (WiFi, 4G, 5G)

There are a lot of advantages to doing things this way. What with light being used, the speed at which the information is transmitted is very, very high up to 100 Gbit/s, in theory; 5 times faster than 5G. Furthermore, the sheer number of LED bulbs already around us hints at a potential future wherein (cheap) access points to receive data are everywhere. Finally, the light waves used as the basis for LiFi do not pass through walls (but can however be reflected off of them). The risk of hacking is therefore much lower than with WiFi, though this seriously limits indoor use cases.

On the other hand, the use of LiFi requires one to be near an operating light source. Its range is thus very limited, and interference is possible with other light sources such as natural sunlight.As of today, this future technology is very much of a niche, despite having been hyped in some circles for half a decade now. One obstacle to popular adoption is the size and price of photoreceptors. As such, key use cases are within areas that are particularly sensitive to hacking and/or electromagnetic interference, such as hospitals, aircraft, military operations

There are also talks of using it for unmanned underwater vehicles, as well as for street lights, to communicate with smart vehicles, or about municipal services to citizens should the technology become more democratised.

3. Energy Storing Bricks

Researchers have managed to store energy in the cheap red bricks, we've been using for construction throughout the world for centuries. This process works for two reasons: bricks are porous and contain something called hematite (which gives it its colour).

For these bricks to store and then release energy, researchers heat them to 160 degrees and vaporize their surface with hydrochloric acid mixed with an organic compound called EDOT. When in contact with hematite, this mix causes a chemical reaction, creating a new plastic nano fiber coating called PEDOT. This polymer is trapped in the porous surface of the brick, forming a continuous and electrically conductive layer on each of its faces. The humble building block can then act as an ion sponge to store and conduct electricity.

For now, the amount of energy these bricks can store is still low, but the proof of concept is a staggering success. It's possible to power a small lamp for 50 minutes with 60 bricks, which doesn't sound like much until you realise it only takes 13 minutes for these bricks to recharge. This future technology also has a long lifespan, since even after 10,000 storage and retrieval cycles, the bricks still retain 90% of their original capacity, without altering the rate of charge and discharge.

The main benefit of this technology of the future would come about when used at a house equipped with solar panels. The bricks could then store unused electricity and thus compensate for the intermittence of this renewable energy.

This would make our homes more self-sufficient in energy and less dependent on electric cables and/or the likes of lithium batteries. Discussions are currently underway with several companies in Europe and the United States to consider its commercialisation and know that the next generation of bricks will be able to increase its energy capacity by 50%. Enough to charge a laptop? Only time will tell.

4. Robotic bees

Let's *bee* honest, it's not great out there, climate-wise. And that's killing bees, which we need to pollinate 35% of our crops, which we need for food, which we need for not being dead. Are we going to stop climate change to save the bees? Of course not, that's not how we roll. Instead, we're going to create robotic bees to pollinate plants just as the real things do. And by "we", I mean Walmart.

Details are scarce, but most researchers estimate that the bees would work by attaching horse hair coated with ionic liquid gel to a tiny drone. The hair picks up pollen from one flower, and moves it to the next. Researchers at Harvard have long been working on "RoboBees" using such techniques. What Walmart offers on top is a wide array of sensors, cameras, artificial intelligence to locate the relevant crops and pollinate them as needed.

If the costs of operating such future technology continues to decrease, we could see autonomous insect pollinate large fields in the coming years, which could save thousands of farmers from ruin, and ensure we can still have almond milk on the superstores' shelves.

I however have a few gripes with it. Firstly, small farmers might never be able to afford robotic bees, and we would once again be empowering the big guy against the little guy. We also don't know how these "bees" would impact the fauna overall, both on and off the fields.

5. Unnamed Dynamic Neural Networks Technology

Neural network uses hidden layers to break down information (the input-images, audio, videos, handwritten text) into tiny pieces of easily understandable components, allowing a computer to inform a prediction about the nature of said input. It does this thanks to a wide array of training data and mathematical models. In doing so, it works "similarly" to our brain, hence the technology's name. This is far from new, but the world of data science has been on the lookout for faster and more efficient ways of using neural networks to serve the upcoming IoT revolution.

First and foremost, the technology reduces computing resources required of the host CPU and cuts back on costs of running data centres, something which seems benign compared to all the issues discussed in this article, but is nevertheless incredibly important at scale. It also reduces the environmental costs of these centres, which is always a plus.

By optimising neural networks for speed, accuracy and power consumption, the technology will lead to improvements in the realm of surveillance, advanced driver assistance systems, autonomous vehicles, vision-guided robotics, drones, augmented and virtual reality, acoustic analysis, and Industrial Internet-of-Things. All worthy endeavours, if you're into that type of thing.

6. Seawater fuel

When it comes to energy, the sea remains a massive untapped resource; it's not like we're going to run out of it anytime soon, what with the North Pole melting. And according to researchers, it could soon be put to good use as fuel for ships. Bear with me: the process uses a potassium promoted molybdenum carbide catalyst to extract carbon dioxide from seawater, turning it into carbon monoxide via the reverse water-gas shift (RWGS) reaction. The carbon monoxide can in turn be converted into a hydrocarbon via the Fischer-Tropsch synthesis Ships can then use this hydrocarbon instead of pre-onboarded fuel which, as we know, carries many risks.

The fuel currently used to power the thousands of ships that cross the seas and oceans daily is very polluting. The conversion of CO2 into chemicals and value-added fuels could significantly reduce the greenhouse gases they emit over the long term.

There are however a handful of issues that need fixing before we get to greener maritime routes. Firstly, carbon dioxide concentration in seawater is about 100 milligrams per litre. That's not much. To put it into perspective, you'd need to process close to 45 million cubic meters of water to power a cruise ship for about a week. And the more water you process, the more sea life you remove from the food chain, with potentially catastrophic long-term results. Secondly, you're still releasing carbon into the air at the end of the day, even if it's water-based carbon. You could argue that it's a net neutral as it will get back to the sea eventually, but if you're at all ecologically conscious, you know this is a slippery slope.

7. **20 minute water**

The recipe was concocted especially for developing countries remote areas where people don't have access to chemical treatments such as chlorine. It could save some of the 300000 children under 5 who die worldwide every year of waterborne diseases such as the cholera, typhoid and hepatitis. It could also help some of the 2.2 billion people who don't have a waste water treatment system.

Unlike other innovations in this article, this one hardly has any downsides. It's very cheap: the amount of silver used for the nano wires is so small the cost is negligible, and the electricity needed can be easily supplied by a small solar panel or a couple of 12-volt car batteries. Since the filter doesn't trap bacteria (killing them instead), it can have much larger pores, allowing water to speed through at a more rapid rate. More than 80,000 times faster than existing filters, to be exact. And it does so without clogging, an issue that plagues existing solutions.

8. Zero-knowledge proof

Privacy: ever heard of it? Computer scientists are perfecting a cryptographic tool we could use to prove something without revealing the information underlying the proof. It sounds incredible but not impossible once you wrap your head around the concept and the fact that it's a bit more complex than saying "c'mon bro, you know I'm good for it".

Allow me to simplify through an example. Imagine, if you will, that a man named John has a blind friend named Jane. He also has in his possession two marbles of different colours, though they are identical in shape and size. Jane puts them behind her back and shows one to John. She then does it again, either changing the marble or showing the same one again, asking if this is the same as the marble previously shown.

If John were guessing whether it was the same or not, he would have a 50/50 chance of getting it right, so she does it again. And again. And because John sees the marbles' colours, he gets it right each time, and the chance that he guessed lucky diminishes. Jane thus knows that John knows which marble is the original shown (and its colour), without her ever knowing the colour of any of the marbles. Boom, zero-knowledge proof. Or zero-knowledge succinct non-interactive argument of knowledge if you're family. Obviously, it all gets mathematical and cryptographic from here, but you get the gist.

It's easy to come up with very cool use cases. For example, if an app needs to know that you have enough money to put a transaction through your bank could communicate that yes, that is the case, without giving an amount. It could also help identify a person without a birth certificate, allow someone to enter a restricted website without needing to display their date of birth... or help with nuclear disarmament. additionally, it could provide proof of a crypto-currency transaction without revealing its amount (BitCoin is, and always will be, too public for me Go Zcash!). Yay for privacy, and here's to a hopefully dying targeted advertising industry.

9. YOLOv5

Real-time object detection is a technique used to detect objects from video. It's the underlying technology behind well, most things we want to use in the future, from Tesla's self-driving cars to Amazon's cashier-less stores. The YOLO ("You Only Look Once") models refer to some of the most versatile and famous real-time object detection and labelling models.

The latest iteration (V5) of the model is worthy of this list for a couple of reasons. It's written in PyTorch, which will make its deployment to mobile a lot easier. It's also quick. Very quick. 140 frames per second quick, while also preserving accuracy. Finally, YOLOv5 is nearly 90 percent smaller than YOLOv4. This means it will be deployed to embedded devices much more easily.

This innovation is important because it means we will soon be able to do real real-time analysis detecting and labelling videos 140 times per second. For context, previous models struggled to get to 10 frames per second. At this speed, you can use AI on videos of fields as varied as the world of medicine or that of sport. It can also improve things like detecting obstacles in autonomous cars earlier to avoid even more collisions. It's a really big deal, but one you surely won't hear about in the newspapers.

10.4D - printing

The name 4D printing can lead to confusion: I am not implying that humanity will be able to create and access another dimension (Only Rubik can do that) Put simply, a 4D-printed product is a 3D-printed object which can change properties when a specific stimulus is applied (submerged underwater, heated, shaken, not stirred). The 4th D is therefore Time; time needed for the stimulus to be relevant.

The key challenge of this technology is obviously finding the relevant "smart material" for all types of uses (namely a hydrogel or a shape memory polymer for the time being). Some work is being done in this space, but we're not even close to being customer-ready, having yet to master reversible changes of certain materials.

The applications are still being discussed, but some very promising industries include healthcare (pills that activate only if the body reaches a certain temperature),fashion (clothes that become tighter in cold temperature), and home-making (furniture that becomes rigid under a certain stimulus).

Another really cool use case is computational folding, wherein objects larger than printers can be printed as only one part.

Obviously, it's always possible that some of these innovations will never become popularised. This should however not stop us from aiming to better understand them: the knowledge gained through planning is crucial to the selection of appropriate actions as future events unfold. A basic understanding of new technologies of the future can also help us steer clear of potentially destructive ones.

References

Basic Concepts of Ecology. IGNOU Website:
Chadha, Kashika & Saha, Kausumi (2020) Digital Literacy in India: Structural Constraints and NEP 2020. Website: https://sprf.in/digital-literacy-in-india-structural-constraints-and- the-nep-2020/

Digital Literacy -Wikipedia the free encyclopedia.Website:
Website: https://en.wikipedia.org/wiki/Digital_literacy

Krish Chetty, Liu Qigui, Nozibele Gcora, Jaya Josie, Li Wenwei, and Chen Fang (2018) Bridging the digital divide: measuring digital literacy. Economics, The Open-Access, Open Assessment E- Journal. Vol. 12, April 25, 2018 | *Google Scholar* Website: http://dx.doi.org/10.5018/economics-ejournal.ja.2018-23.

Fillmore, John (2021) Why Digital Upskilling Is A Crucial part of Sustainability. 16 Dec. 2021.
The Digital Economy. Website:
https://www.weforum.org/agenda/2021/12/digital- upskilling-sustainability/

Gore, Vittal (2013) 21st Century Skills and Prospective Job Challenges. **IUP Journal of SoftSkills; Hyderabad** Vol.7,Iss.4, (Dec 2013): 7-14 Google Scholar Website.

Gonda, Rob (2018) Adaptibility is Key to Survival In The Age of Digital Darwinism. Forbes May 24, 2018.
Website :https://www.forbes.com/sites/forbestechcouncil/2018/05/24/adaptability-is-key-to-survival-in-the-age-of-digitadarwinism/?sh=11407d7e408c

Laal, Marjan (2011) Lifelong Learning: What does it Mean? Procedia - Social and Behavioral Sciences 28 (2011) 470 – 474.
Website:https://www.sciencedirect.com/science/article/pii/S1877042811025298
Muthkoor, V. and Mumtaz, F (2021) The Digital Dream: Upskilling India For The Future.
Website: https://www.ideasforindia.in/topics/governance/the-digital-dream-upskilling-india-for-the-future.html

MacMinn, James (2019) What are Some New Digital Technologies? – Blog. MatchBox. 12,
November 2019 Website: What Are The New Digital Technologies | 8 New Digital Technologies (matchboxdesigngroup.com)

New Age Technology and Digitization- Website.Website: New Age Technology & Digitization (foxmandal.global)

O'Donnell, Jim (2020) Digital Transformation Requires Upskilling and Reskilling Website:02, October, 2020

Website:https://www.techtarget.com/searcherp/feature/Digital-transformation-requires-upskilling-and-reskilling

OECD (2007, April). Qualifications and Lifelong Learning (p.1), Organization for Economic
Co-operation and Development, Policy brief. Retrieved 15 May 2011 from: www.oecd.org/dataoecd/10/2/38500491.pdf

Stowers, Sean (2022) Improve Your Workforce's Digital Literacy with Upskilling. Training Industry. Website: https://trainingindustry.com/articles/workforce-development/improve-your-workforces-digital-literacy-with-upskilling/

Technostress- Wikipedia, the free encyclopedia
Website: https://en.wikipedia.org/wiki/Technostress

The Importance of Digital Literacy in Today's World (exploringyourmind.com)
Website:https://exploringyourmind.com/the-importance-of-digital-literacy-in-todays-world/

Never Underestimate the Power of Human Adaptability.
Website:https://www.cognizant.com/futureofwork/article/never-underestimate-the-power-of-human-adaptability. Nov 01, 2018

UNESCO Institute for Information Technologies in Education (2011). Digital Literacy in Education. Policy Brief, May 2011.
Website: http://unesdoc.unesco.org/images/0021/002144/214485e.pdf

UNESCO (United Nations Educational Scientific and Cultural Organization) (2016). Global Education Monitoring Report. Education for People and Planet: Creating Sustainable Futures for All. Website: http://unesdoc.unesco.org/images/0024/002457/245752e.pdf.

Department of School Education & Literacy, MHRD, Government of India (June 2020) India Report Digital Education Remote Learning Initiatives across India, New Delhi.

National Council of Educational Research and Training, Ministry of Education Report of the Education Commission, 1964-66, New Delhi.

Ministry of Law & Justice, Government of India The Right of Children to Free and Compulsory Education Act, 2009, New Delhi.

Annual Status of Education Report [ASER]. (2018). Annual status of education report (rural) 2017. Retrieved from
http://img.asercentre.org/docs/Publications/ASER%20Reports/ASER%202017/aser2017fullreportfinal.pdf.

Bansal, S.(2016). Teacher education programmes preparing teachers for inclusive classrooms: A North India context. *Journal of Disability Management Rehabilitation*, 2(2), 83–90.

Bawane, J. (2019). Paradoxes in teacher education: Voices from the Indian context. In R. Setty, R. Iyengar, M. A. Witenstein, E. J. Byker, & H. Kidwai (Eds.), *Teaching and teacher education. South Asian perspectives* (pp. 49–70). Cham: Springer.

Bhatnagar, N., & Das, A. (2013). Attitudes of secondary school teachers towards inclusive education in New Delhi, India. *Journal of Research in Special Educational Needs*. doi:10.1111/j.1471-3802.2012.01262.

Das, A., & Kattumuri, R. (2011). Children with disabilities in private inclusive schools in Mumbai: Experiences and challenges. *Electronic Journal for Inclusive Education*,

2(8).https://corescholar.libraries.wright.edu/cgi/viewcontent.cgi?article=1136&context=ejie.

Das, A., Kuyini, A. B., & Desai, I. (2013). Inclusive education in India: Are the teachers prepared? *International Journal of Special Education*, 28(1), 27–36.

District Primary Education Programme (2000). *Empowerment through education: Identification and enrolment of children with special needs in DPEP*. Noida, India: Education Consultant of India Limited.

Education Commission. (2019). Transforming the education workforce: Learning teams for a learning generation. Retrieved from https://educationcommission.org/wp-content/uploads/2019/09/Transforming-the-Education-Workforce-Full-Report.pdf

Jha, M. M. (2010). From special to inclusive education in India. Case studies of three schools in Delhi. India: Pearson Education.

Kalita, U. (2017). A study on attitude of primary school teachers towards inclusive education. International Journal of Advanced Education, 2(3), 127–130

Kumari, P., Nayan, R., Aggarwal, S. P., & Baswani, G. (2019). Rethinking teacher education programmes for inclusive classrooms: Issues and challenges in India. International Journal of Information Education Technology, 9(2), 143–148.

Ministry of Human Resource Development [MHRD]. (2011). Sarva Shikshya Abhiyan. Framework for Implementation. Retrieved from http://ssa.nic.in/page_portletlinks?foldername=ssa-framework

Ministry of Human Resource Development [MHRD]. (2018). Samagra Shiksha Abhiyan: An integrated scheme for school education. Framework for implementation. Draft Document. Retrieved from http://samagra.mhrd.gov.in/docs/Framework_IISE%20_F.pdf .

MLJ. (1995). The persons with disabilities (equal opportunities, protection of rights and full participation) act, 1995. New Delhi: Government of India. Retrieved from http://socialjustice.nic.in/pwdact1995.php

Ministry of Law and Justice [MLJ]. (2009). Right to education act. Retrieved from http://mhrd.gov.in/acts_rules_SE

Ministry of Law and Justice [MLJ]. (2016). The rights of persons with disabilities act, 2016. New Delhi: Government of India Retrieved from http://socialjustice.nic.in/pwd2011.php.

Mukhopadhyay, S., & Mani, M. N. G. (2002). Education of children with special needs: A profile of basic education. In R. Govinda (Ed.), India education report (pp.96–108). New Delhi Oxford University Press.

National Council for Teacher Education [NCTE]. (2009). National curriculum framework for teacher education. Towards preparing professional and humane teacher. New Delhi: Author.

National Council for Teacher Education [NCTE]. (2014). Curriculum framework for 2-year B.Ed. New Delhi: Author.

Neuman, W. L. (2014). Social research methods :Qualitative and quantitative approaches. Essex, UK: Pearson.

Shah, R., Das, A., Desai, I., & Tiwari, A. (2016). Teachers' concerns about inclusive education in Ahmedabad, India. Journal of Research in Special Educational Needs, 16(1), 34–45.

Sharma, U., Moore, D., & Sonawane, S. (2009). Attitudes and concerns of pre-service teachers regarding inclusion of students with disabilities into regular schools in Pune, India. Asia-Pacific Journal of Teacher Education, 37(3), 319–331.

Singal, N. (2008). Working towards inclusion: Reflections from the classroom. Teaching and Teacher Education: An International Journal of Research and Studies, 24(6), 1516–1529.

Singal, N. (2016). Schooling children with disabilities: Parental perceptions and experiences. International Journal of Educational Development, 50, 33–40.

Singal, N. (2019). Challenges and opportunities in efforts towards inclusive education: Reflections from India. International Journal of Inclusive Education, 23(7–8), 827–840.

Singal, N., De, A., & Bhutani, S. K. (2016). A critical analysis of the education of children with disabilities under RMSA: Resourcing and provision. Report prepared for Cambridge Education

U-Net: Convolutional Networks for BiomedicalImage Segmentation: Olaf Ronneberger, Philipp Fischer, and Thomas Brox, 2015

Brain Tumor Segmentation and RadiomicsSurvival Prediction: Contribution to the BRATS2017 Challenge: Fabian Isensee, Philipp Kickingereder, Wolfgang Wick, Martin Bendszus,and Klaus H. Maier-Hein.

nnU-Net for Brain Tumor Segmentation : Fabian Isensee, Paul F. Jager, Peter M. Full, Philipp, and KlausH.Maier-Hein.

KiU-Net: OvercompleteConvolutionalArchitectures for Biomedical Image andVolumetric Segmentation: Jeya Maria Jose Valanarasu,Student Member, IEEE, Vishwanath A. Sindagi,Student Member, IEEE,IlkerHacihaliloglu,Member, IEEE,and Vishal M. Patel,Senior Member, IEEE

DoubleU-Net: A Deep Convolutional NeuralNetwork for Medical Image Segmentation : Debesh Jha, Michael A. Riegler, Dag Johansen [27 Jun 2020] [Double U-Net (10-13)]

Bi-Directional ConvLSTM U-Net with Densley Connected Convolutions: Reza Azad, Maryam Asadi-Aghbolaghi, Mahmood Fathy [31 Aug 2019] [BCDU-Net (14-17)]

https://towardsdatascience.com/illustrated-guide-to-lstms-and-gru-s-a-step-by-step-explanation-44e9eb85bf21.

[Bi-directional Conv LSTM(18)]

https://towardsdatascience.com/understanding-and-visualizing-densenets-7f688092391a. [LSTM(19)]

https://www.sciencedirect.com/topics/computer-science/image-segmentation . [image segmentation(6)]

UNet++: A Nested U-Net Architecturefor Medical Image Segmentation: Zongwei Zhou, MdMahfuzur Rahman Siddiquee,NimaTajbakhsh, and Jianming Liang.[18 Jul 2018]

SCAU-Net: Spatial-Channel Attention U-Net for Gland Segmentation: Peng Zhao, Jindi Zhang, Weijia Fang and Shuiguang Deng [ORIGINAL RESEARCH published: 03 July 2020doi: 10.3389/fbioe.2020.00670]

Building Detection on Aerial Images Using U-NET Neural Networks: Leonid Ivanovsky, Vladimir Khryashchev, Vladimir Pavlov , Anna Ostrovskaya.[ISSN 2305-7254]

A Probabilistic U-Net for Segmentation of Ambiguous Images: Simon A. A. Kohli, Bernardino Romera-Paredes, Clemens Meyer, Jeffrey De Fauw, Joseph R. Ledsam, Klaus H. Maier-Hein,S. M. Ali Eslami, Danilo Jimenez Rezende, and Olaf Ronneberger. [NeurIPS 2018]

U^2-Net: Going Deeper with Nested U-Structure for Salient Object Detection:Xuebin Qin, Zichen Zhang, Chenyang Huang, Masood Dehghan, Osmar R. Zaiane, Martin Jagersand[2020]

UNet 3+: A Full-Scale Connected UNet for Medical Image Segmentation: Huimin Huang, Lanfen Lin, Ruofeng Tong, Hongjie Hu, Qiaowei Zhang, Yutaro Iwamoto, Xianhua Han, Yen-Wei Chen, Jian Wu [2020]

Abadi, M., Barham, P., Chen, J., Chen, Z., Davis, A., Dean, J., Devin, M., Ghemawat, S., Irving, G., Isard, M., Kudlur, M., Levenberg, J., Monga, R., Moore, S.,Murray, D.G., Steiner, B., Tucker, P., Vasudevan, V., Warden, P., Wicke, M., Yu,Y., Zheng, X., Brain, G.: TensorFlow: A system for large-scale machine learning.In: OSDI. pp. 265–284 (2016)

Abdulkadir, A., Lienkamp, S.S., Brox, T., Ronneberger, O.: 3D U-Net :Learningdense volumetric segmentation from sparse annotation. In: MICCAI. pp. 424–432(2016)

Bakas, S., Akbari, H., Sotiras, A., Bilello, M., Rozycki, M., Kirby, J., Freymann,J., Farahani, K., Davatzikos, C.: Segmentation labels and radiomic features for thepre-operative scans of the TCGA-LGG collection. The Cancer Imaging Archive(2017)

A. Is ̧ın, C. Direko ̆glu, and M. S ̧ ah, "Review of MRI-based brain tumorimage segmentation using deep learning methods,"ProcediaComput.Sci., vol. 102, pp. 317–324, 2016.

B. H. Menze, A. Jakab, S. Bauer, J. Kalpathy-Cramer, K. Farahani,J. Kirby, Y. Burren, N. Porz, J. Slotboom, R. Wiestet al., "Themultimodal brain tumor image segmentation benchmark (BRATS),"IEEE Trans. Med. Imag., vol. 34, no. 10, pp. 1993–2024, 2015

S. Bauer, R. Wiest, L.-P. Nolte, and M. Reyes, "A survey of MRI-based medical image analysis for brain tumor studies,"Phys. Med. Biol.,vol. 58, no. 13, pp. R97–R129, 2013.

S. Pereira, A. Pinto, V. Alves, and C. A. Silva, "Brain tumor segmentation using convolutional neural networks in MRI images,"IEEETrans.Med. Imag., vol. 35, no. 5, pp. 1240–1251, 2016

M. Havaei, A. Davy, D. Warde-Farley, A. Biard, A. Courville, Y. Bengio,C. Pal, P.-M. Jodoin, and H. Larochelle, "Brain tumor segmentation withdeep neural networks,"Med. Imag. Anal., vol. 35, pp. 18–31, 2017.

Image Segmentation Algorithms Overview: Song Yuheng,YanHao

Deep Learning Techniques for Medical Image Segmentation:Achievements and Challenges: Mohammad Hesam Hesamian1,Wenjing Jia1,Xiangjian He,Paul Kennedy

MachineLearningTechniquesforBiomedicalImageSegmentation:AnOverviewofTechnicalAspectsandIntroductiontoState-of-ArtApplication

HyunseokSeo,MasoudBadieiKhuzan,VarunVasudevan,CharlesHuang,HongyiRen, RuoxiuXiao,XiaoJia,andLeiXing

Medical Image Segmentation Using Deep Learning:A Survey: Tao Lei,Risheng Wang, Yong Wan, Bingtao Zhang, HongyingMeng

Visit our website for more details

www.ppdiinstitute.in

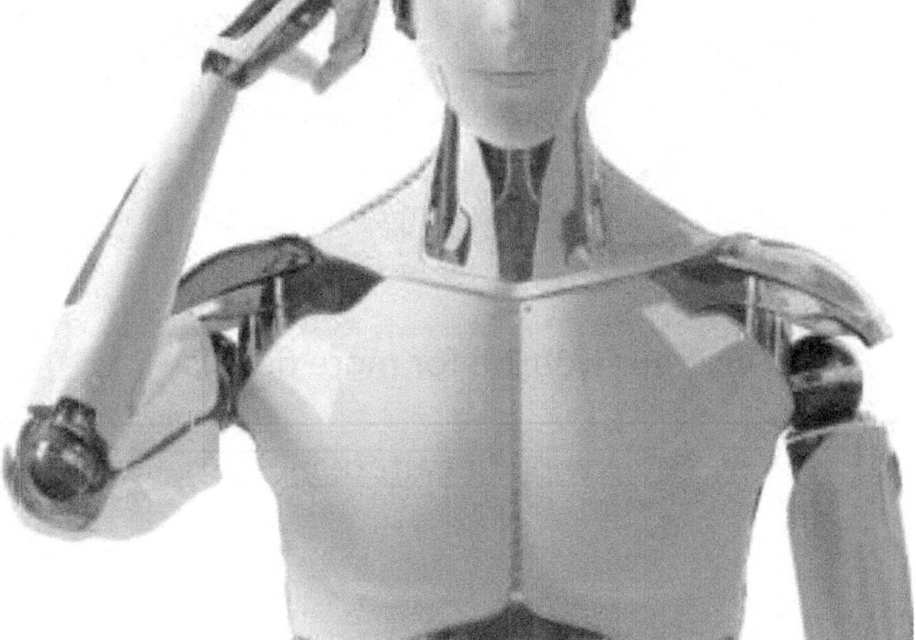